T0248665

Health and Sustainable Agricultural Development

Health and Sustainable Agricultural Development

Perspectives on Growth and Constraints

EDITED BY
Vernon W. Ruttan

Routledge
Taylor & Francis Group

LONDON AND NEW YORK

First published 1994 by Westview Press, Inc.

Published 2018 by Routledge
52 Vanderbilt Avenue, New York, NY 10017
2 Park Square, Milton Park, Abingdon, Oxon OX14 4RN

Routledge is an imprint of the Taylor & Francis Group, an informa business

Copyright © 1994 Taylor & Francis

All rights reserved. No part of this book may be reprinted or reproduced or utilised in any form or by any electronic, mechanical, or other means, now known or hereafter invented, including photocopying and recording, or in any information storage or retrieval system, without permission in writing from the publishers.

Notice:
Product or corporate names may be trademarks or registered trademarks, and are used only for identification and explanation without intent to infringe.

Library of Congress Cataloging-in-Publication Data
Health and sustainable agricultural development: perspectives on
growth and constraints / edited by Vernon W. Ruttan.
    p. cm.
Includes bibliographical references.
ISBN 0-8133-8838-4
    1. Public health—Developing countries. 2. Nutrition—Developing
countries. 3. Agriculture—Health aspects—Developing countries.
4. Medical policy—Developing countries. 5. Sustainable
development—Health aspects—Developing countries. 6. Sustainable
agriculture—Health aspects—Developing countries. 7. Tropical
medicine. I. Ruttan, Vernon W.
RA441.5.H4267 1994
362.1'09172'4—dc20   94-11363
                                                                          CIP

ISBN 13: 978-0-367-01782-8 (hbk)

ISBN 13: 978-0-367-16769-1 (pbk)

Contents

PART SIX
PERSPECTIVES AND REFLECTIONS

Preface

This book is the product of one of a series of consultations held at the University of Minnesota Hubert H. Humphrey Institute of Public Affairs. The participants in the series were a number of leading agricultural, environmental, health and social scientists who were asked to identify the implications of global change for agricultural research priorities into the twenty-first century.

The idea of the consultations emerged out of a series of conversations in the spring of 1988 between Robert W. Herdt, Director for Agricultural Sciences at the Rockefeller Foundation, and the editor. The topic of the discussions was how to focus attention on emerging agricultural research priorities in developed and developing economies in response to the demands they would be placing on their agricultural systems and on their agricultural scientists as we move into the second quarter of the twenty-first century.

The consultations were organized around three broad subject areas: (a) Scientific and Technical Constraints on Crop and Animal Productivity; (b) Resource and Environmental Constraints in Sustainable Growth in Agricultural Production; and (c) Health Constraints in Agricultural Development. They were structured as informal dialogues. No formal papers were requested or presented. The dialogues were taped, and the transcripts were reproduced initially as Staff Papers at the University of Minnesota Department of Agricultural and Applied Economics.

This publication is based on the third of the series. Copies of the first can be obtained from Waite Library, Department of Agricultural and Applied Economics, University of Minnesota, 1994 Buford Avenue, St. Paul, MN 55108. For the second consultation, see Vernon W. Ruttan (ed.) Sustainable Agriculture and the Environment: Perspectives on Growth and Constraints. (Boulder: Westview, 1992)

Vernon W. Ruttan

ABOUT THE CONTRIBUTORS

C. Eugene Allen is Vice President with responsibility for Agriculture, Natural Resources and Home Economics at the University of Minnesota. He has been Dean of the College of Agriculture and Associate Director of the Agricultural Experiment Station at the University of Minnesota since 1984. Previously, he was a faculty member in the Departments of Animal Science and Food Science and Nutrition. His research on animal growth biology and the functional and nutritional characteristics of animal food products has earned him numerous awards from professional societies. He has served on the Board on Agriculture of the National Research Council of the National Academy of Sciences.

David E. Bell is Clarence Gamble Professor of Population Sciences and International Health, Emeritus, at the Harvard School of Public Health. Before joining Harvard in 1981, he was Vice President for international programs at the Ford Foundation (1966-1981), Administrator, U.S. Agency for International Development (1962-66), and Director, U.S. Bureau of the Budget (1961-62). He is an economist with long-standing interests in institutional development, beginning when he was the field leader of a team of advisors (1954-57) helping the Pakistan Planning Board prepare that country's first five-year development plan. During 1987-90 he served as Senior Consultant to the International Commission on Health Research for Development.

Zbigniew Bochniarz is visiting Professor/Senior Fellow in the Hubert H. Humphrey Institute of Public Affairs, University of Minnesota. He is on leave from the Warsaw School of Economics, Poland. His research interests are in the fields of environmental problems, institutional design for sustainable development, and economic transition in Central and Eastern Europe. His published works include *Industry in Service for Sustainable Development (1992), Institutional Challenge for Sustainable Development in Czechoslovakia (1992)*, and *Designing Institutions for Sustainable Development: A New Challenge for Poland (1991)*. Since 1991, he has been a Director of two multi-year USAID sponsored projects oriented towards institutional and human capacity building for sustainable development in CEE.

David Bradley is Professor of Tropical Hygiene at the London School of Hygiene and Tropical Medicine, in the Department of Epidemiology and Population Sciences and the Ross Institute. He is a physician and biologist. He lived for 10 years in East Africa, working on schistosomiasis and on domestic water supply policy in relation to disease; then for several years in Oxford on the genetic determinants of susceptibility to leishmaniasis and other infections. His recent work has been on malaria epidemiology and environmental change. He was one of the authors of *Drawers of Water* (1972), *Sanitation and Disease* (1983) and is editor of the *Journal of Tropical Medicine and Hygiene.* He has for many years been closely involved with WHO programs on tropical diseases, diarrhoeal diseases, and the building of research capacity in developing countries.

Doris H. Calloway is Emerita Professor of Nutrition in the Department of Nutritional Science at the University of California (Berkeley). She received her B.S. on nutrition and dietetics from Ohio State University and her Ph.D. from the University of Chicago. She has served as Chair of the Department of Nutritional Sciences and Provost at the University of California at Berkeley. Her research has been in the area of nutrition and health. During 1981-87 she chaired a major AID funded International Collaborative Research Program on nutrition and human function. She is a Fellow of the American Institute of Nutrition and a member of the Institute of Medicine.

James Chin is Clinical Professor of Epidemiology at the School of Public Health, University of California at Berkeley, and an international consultant on the impact of AIDS in developing countries. He was Chief of the Surveillance, Forecasting, and Impact Assessment Unit, of the Global Programme on AIDS, World Health Organization, Geneva, Switzerland from 1987 to 1992. Prior to joining WHO he was the State Epidemiologist, responsible for infectious disease control in the California State Department of Health Services in Berkeley, California from 1968-1987.

Donald Henderson is Deputy Assistant Secretary for Health Science in the Department of Health and Human Services. Previously, he served as Associate Director for Life Sciences in the Office of Science and Technology Policy, Executive Office of the President (1991-1993), Dean and Professor of Epidemiology and International Health, the Johns Hopkins University School of Hygiene and Public Health (1977-1991) and Director of the World Health Organization's

Smallpox Eradication Program (1966-1977). He is an epidemiologist with long-standing interests in international health and prevention.

Robert Herdt, Director for Agricultural Sciences at the Rockefeller Foundation since 1987, has actively worked with biological and social scientists over the past two decades to improve production and income of small-scale farmers in developing countries. He has headed the Economics Department of the International Rice Research Institute and acted as Scientific Advisor to the CGIAR Secretariat at the World Bank. He wrote *The Rice Economy* (with Randolph Barker) and *Science and Food: The CGIAR and Its Partners* (with Jock Anderson and Grant Scobie).

Bruce Johnston is Professor Emeritus in the Food Research Institute, Stanford University. He is an economist with long-standing interests in the role of agriculture in economic development. He was co-editor of the SSRC's symposium volume on Agricultural Development and Economic Growth (1967) and co-author with Peter Kilby of Agriculture and Structural Transformation: Economic Strategies in Late-Developing Countries (1975). His 1982 book with William C. Clark, Redesigning Rural Development, and a forthcoming book with Thomas P. Tomich and Peter Kilby, Transforming Agrarian Economics: Opportunities Seized, Opportunities Missed give considerable attention to health and family planning programs as elements in well-conceived development strategies.

John Murray is Professor Emeritus in the Department of Medicine at the University of Minnesota. He has studied the relationships among nutrition, disease, and health in a number of Sahelian countries in Africa.

Michael Osterholm is an Epidemiologist in the Minnesota State Department of Health and Adjunct Associate Professor in the Department of Epidemiology at the University of Minnesota.

Mark Rosenzweig is Professor of Economics and a Research Associate in the Populations Studies Center at the University of Pennsylvania. He was formerly a Professor in the Department of Economics at the University of Minnesota, where he was also co-Director of the Economic Development Center. His research has covered the fields of population growth, labor markets, family behavior and human capital formation, with particular attention to the health of children, in both developing and developed countries. His most recent work has focussed on evaluating the effects of the growth in health and family planning programs in Indonesia and on assessing the impact of agricultural development on health and population change in India.

Vernon Ruttan is a Regents' Professor of Economics and Agricultural Economics and an Adjunct Professor in the Hubert H. Humphrey Institute of Public Affairs at the University of Minnesota. His research has been in the field of agricultural development, resource economics, and research policy. Ruttan is the author of Agricultural Research Policy (University of Minnesota Press, 1982); (with Yujiro Hayami) Agricultural Development: An International Perspective (Johns Hopkins Press, rev. ed. 1985); and Agriculture Environment and Health: Sustainable Development in the 21st Century (University of Minnesota, 1993). Ruttan has been elected a fellow of the American Academy of Arts and Sciences (1976), the American Association for the Advancement of Science (1986), and to membership in the National Academy of Sciences (1990).

John Sanders is a Professor in the Department of Agricultural Economics at Purdue University. He worked in Latin America and Portugal for over a decade before coming to Purdue in 1983. His principal research interest is in the economics of technological change in agriculture. He has been working on this topic in the Sahel and other regions of Sub-Saharan Africa for the last decade. He is the co-author of The Economics of Agricultural Technology Development in Semi-Arid Sub-Saharan Africa (Baltimore: The Johns Hopkins University Press, forthcoming).

Conrad Straub is Professor Emeritus, Environmental Health, School of Public Health, University of Minnesota. Before joining the University in 1966, he served with the US Public Health Service for 25 years in foreign and domestic field and research assignments. He was a consultant to the World Health Organization in the areas of radiological health and environmental engineering with assignments in Poland, Hungary, Pakistan, and the Pacific countries. He was co-editor and editor of a series of Environmental Control Handbooks, and edited the Journal Critical Reviews in Environmental Control from 1969-1972.

Kenneth S. Warren is Vice President for Academic Affairs at the Picower Institute for Medical Research. His previous positions include Director for Science of the Maxwell Communication Corporation, Director for Health Sciences of the Rockefeller Foundation, and Professor of Medicine and of Library Sciences at Case Western Reserve University School of Medicine. His specialty is tropical medicine, and in addition to his work in the laboratory and clinic, he has done field research in Brazil, St. Lucia, Kenya, Egypt, Thailand, the Philippines and China. He is a member of the Institute of Medicine of the National Academy of Science. His publications include *Tropical and Geographical Medicine*, McGraw Hill, 1990 and *Immunology and Molecular Biology of Parasitic Infections*, Blackwell Scientific Publications, 1993.

Thomas Whitmore is Assistant Professor of Geography at the University of North Carolina. His research centers on the analysis of agriculture and demography in Mexico at the time of European contact. He is currently co-authoring a book examining Amerindian agriculture in Middle America in 1492. His published works include *Disease and Death in Early Colonial Mexico: Simulating Amerindian Depopulation* (1992).

PART ONE

Introduction

1

Concerns About Health and Sustainable Agricultural Development

Vernon W. Ruttan

In this book a group of leading health, agricultural and social scientists explore the role of health in sustainable agricultural development into the early decades of the twenty-first century.

The conversations reproduced here were conducted over a period of two days during the summer of 1990. The conversations were the third in a series of three consultations that began in the late 1980s. (Ruttan, 1989, 1992) The objective of the consultations was to explore the implications of global change for agricultural research priorities and organization as we move into the early decades of the 21st Century.

The first consultation dealt with the biological and technological constraints on crop and animal production. It may seem a bit strange, after a decade of enthusiastic publicity about the promise of biotechnology, to be very concerned about agricultural productivity. Since the early 1980s there has been more concern about surpluses than shortages. But there are indications that the advances in science and technology that were responsible for the rapid growth in the 1960s and 1970s are slowing down. Rice yields in maximum yield trials at the International Rice Research Institute have not risen in 20 years. Farmer's yields are approaching experiment station yields. In the United States corn yields have been going up about two bushels a year since before World War II. But a two bushel increase is a much smaller percentage gain today, when the average yield is over 120 bushels per acre, than it was 50 years ago when the average yield was closer to 30 bushels per acre. What will come next? What will be the encore to the green revolution?

The second consultation dealt with environmental constraints on agricultural development. We were concerned with a set of micro-level constraints and a set of macro-level constraints. The micro environmental constraints are largely the result of agricultural intensification - things like groundwater pollution, erosion, and salinization. The macro environmental constraints are largely the product of industrial intensification - things like acid rain, ozone depletion and global warming.

The topic of this third consultation is a bit more speculative. When one starts asking what would be the impact if a number of health concerns - such as malaria and tuberculosis resurgence; failure to make progress on parasitic disease (with a few exceptions); the failure to use the knowledge we have about infectious disease; the emergence of new viral diseases such as AIDS; and the health effects of environmental change - should come together at the same time it is not too difficult to visualize large numbers of sick people in many villages around the world. The numbers could be large enough to seriously impinge on food production capacity.

The conversations reproduced in this book are organized around five topics. The first establishes the health policy context for subsequent discussion. David E. Bell has distilled some of the major issues that emerged during his work with the Commission on Health Research for Development (1990). Donald Henderson outlines his concerns about the eroding capacity for research on health problems in the tropics and on the failure to organize health services to address issues of health maintenance rather than sickness recovery. A major question that emerged out of these conversations was why the international community has failed to support a system of international health research and the essential national health research in the developing countries of the tropics.

The second topic focused on is the current status and understanding of disease in the tropics. David Bradley draws on his own broad experience and scholarship on environment and parasite disease. Ken Warren talks about the biological and economic constraints on the development of vaccines and drugs for the control of infections caused by bacteria and viruses. James Chin outlines our current understanding of the issues associated with the diffusion of the AID virus, particularly in Africa. A clear implication of these discussions is that it is not beyond the realm of possibility to anticipate the emergence of a global health crisis sometime in the early decade of the next century.

A third conversation focused on the issue of nutrition and environmental health. Michael Osterholm presents an epidemiological perspective on our inadequate understanding and surveillance of the food supply. John Murray draws on his extensive investigation in Africa to outline some of the complex interactions between nutrition and disease. Conrad Straub focuses on the importance of rural water supply for rural health. Zhigniew Bochniariz examines the implications of environmental change in Central and Eastern Europe. It seems clear that the low intensity generic nature of the sources of nutritional and environmental threats to health have contributed to both inadequate surveillance and inadequate programs to address the sources and impacts.

A fourth session focused more directly on the implications of health for agricultural development. Tom Whitmore draws on his historical research on disease-induced population collapse, to provide a perspective on contemporary health problems in developing countries. John Sanders drew on his research on technical change and agricultural production in West Africa to explore the interaction between environment, disease and agricultural production. Mark Rosenzweig drew on his research, particularly in India, to explore the reciprocal relationships between agricultural development and health -- does agricultural development effect health and does health effect agricultural development? Bruce Johnston discusses the relationships between health, population growth and agricultural development. He gives particular attention to the relationship between child survival and birth rate decline. It is not too difficult to anticipate that in may villages in the developing world health will become an even more serious constraint on agricultural development.

In a final session the participants in the consultation attempted to place the conversations of the previous day and a half in perspective. David Bradley lead a discussion on the issue of the tropical disease and work. Donald Henderson and David Bell returned to the issue of the design of health research systems in developing countries. The specific research implications and priorities that emerged from the conversations were then outlined by Vernon Ruttan. It is clear that more effective bridges must be built between the agricultural and health communities.

In a final chapter, David E. Bell and Vernon W. Ruttan outline and contrast the development and design of national and international agricultural research systems to meet the needs of developing countries. They characterized the agricultural and health research systems as "island empires" in need of more effective bridges. They also outline the need for more effective articulation between international research, national research and the communities and families that must be empowered to make more effective use of the knowledge, technology and materials generated by the health and agricultural research systems.

PART TWO

Health Policy and Health Systems

2

A Perspective from the Commission on Health Research for Development

David E. Bell

Bell: In his invitation to participants, Ruttan noted that "health issues could seriously erode production capacity in rural areas." This is an extraordinarily narrow agriculturally centered way of identifying what we're here to talk about. While this is a legitimate and sensible questions it seems to be built around the concept that the purpose of good health is to have healthy workers. Anybody who has spent any time in the health field would reject any such narrow conception. I don't think he meant to be narrow. In a proposal some years ago to the AID Research Advisory Committee (RAC) Ruttan talked about reciprocal relationships. That is what we ought to be concerned with! There are very significant impacts on agricultural production from poor health. At the same time, there are obviously important influences of agricultural development on health. It is this interactive process that we really ought to be concerned about. I also have a second objection to the agenda. What does agricultural production do to the environment? How does it influence the environment in a way that has health effects? He seems to dismiss at the beginning what seems to be the natural and most powerful inter-relationship between health and agricultural production. That is food supply! Nutrition is at the basis of human health. A central question is whether the food supply is going to keep up with another doubling of the world's population over the next half century. We are at five billion plus now and we're probably headed for over ten. How do we respond to this very serious problem?

The issue goes beyond general or global food availability. There are special problems in Africa. There are other special cases that are not geographically based. There are populations in most countries -

including the United States - where the issues is food entitlement or poverty.

One of the things I was educated about by my recent work with the Commission on Health Research for Development (1990) is that there are significant micro-nutrient problems. Deficiencies in Vitamin A have enormous consequences for children. Vitamin A is all around us in food. But studies, such as Sommer's work in Sumatra, indicate that families do not use what is right at hand. It is an extremely significant issue. The infectious and parasitic diseases do have significant impacts on peoples' capacity to be productive in agriculture. I take it the first question this group will need to be focusing on is how important are these effects? Some parasitic diseases like malaria and schistosomiasis affect millions of people. Others affect only a handful. For purposes of trying to relate health and agriculture, presumably the former are much more significant than the latter. We don't need a complete catalogue of all the parasitic diseases to identify those that seem to be enormously significant for agriculture. The number of persons infected is clearly not the right way to measure relative significance. We need some kind of an index of impact. A lot of cases of malaria in a population that has acquired a reasonable degree of immunity may not be very significant in terms of its effect on agricultural production. The question is how do we decide how to focus attention on these parasitic diseases that really make a difference?

A second question is what are the prospects for doing anything about it? There is a sense in which malaria at the moment is out of control. Twenty years ago we had a malaria eradication plan. It turned out to be infeasible. In retrospect we set ourselves back badly by focusing our efforts on eradication. The mosquitos that survived were resistant to pesticides. So what do we do about malaria today? The most promising things that we heard about malaria during the Commission's work were very localized, site specific, community based activities. There are some experiments in Sri Lanka, some in India and some, no doubt, elsewhere. They aren't intended to eradicate malaria. They are intended to reduce its impact in particular communities. It is one potential way of dealing with malaria. It isn't satisfactory from the point of view of health. But it may be reasonably satisfactory in maintaining agricultural or industrial production.

When we turn to the health implications of environmental change we also have to deal with the issue of relative significance. My limited understanding is that heavy use of pesticide and fertilizer are very important health threats. There is research that suggests that as yet there is very little actual health hazard arising from fertilizer use in developing countries (Conway and Petty, 1988). But certainly the use of pesticides and other intensive agricultural technologies does raise very significant health concerns.

Deforestation and soil erosion are important threats to community economic viability in some areas. This is not necessarily the result of modernization of agriculture. It may occur as a result of pushing traditional agriculture farther onto fragile environments.

The main thing the Commission on Health Research for Development identified, and tried to press, as a guideline for action, is the enormous significance of local research capacity. Not fancy laboratory and hospital based capacity, but the ability to identify health problems at the local level. The need is to be able to draw on what is know that will be locally useful. I've already cited the malaria example. The main conclusion was the significance of local research capacity, and the very serious extent to which that is not present in the health field. Consequently, the principal recommendation for action is to develop the capacity to conduct what the Commission called "essential national health research."

In the area of environmental consequences exactly the same point should be made. Where in India, where in Malaysia, where in Nigeria is there serious capacity to address the question of what pesticides are doing to farmers and to children? And what can we do about it? What is the trade-off between the positive contribution of pesticides to food supply and the negative impact on health? There are additional issues that should be on your agenda. One, again from the Commission's findings, is the very high accident rate in agriculture. This isn't simply the pesticide issue. It is people getting run over by tractors and other equipment. It is trying to handle lead in batteries that people don't know anything about. The question of occupational health in agriculture may be even more significant in developing countries than in industrialized ones.

Finally, we must ask much more seriously than in the past, why there is so little successful research and analysis on health and agriculture? Repeatedly, as you can tell from what I have been saying, one wants to see the health and agricultural aspects considered jointly. What do you do in irrigation planning to effect the possibilities of controlling malaria? How do you get at those schistosomiasis-carrying snails? What do you do about pesticides and health? These are joint problems, not separate problems. And yet I don't recall encountering many cases of health researchers and agricultural researchers working effectively together.

Ruttan: Thanks, David. One of the things that really intrigues me is this issue of location-specific or site-specific research capacity. I remember after working at the IRRI in Southeast Asia, in the 1960s, I became very aware of that issue in agricultural research. When I tried to advance the same perspective in the field of health research when I served on the AID Research Advisory Committee the typical response was that we can do all the research in the developed countries -- all they have to do in the developing countries is get out and apply it. I'm really pleased to see that this issue was addressed by the Commission.

Bell: It is important that we finally confront the question - is there a reasonably close analogy between what has been learned in agriculture about the importance of local adaptative research? The Commission arrived at the conclusion independently. It wasn't saying "this is working in agriculture, we ought to do it in health." The person on the Commission who was most influential in laying out this paradigm was Adetokunbo O. Lucas from Nigeria. The Lucas paradigm involves four levels of research, the first being the identification and analysis of the problem. The second is policy analysis of program design and testing. He regards these first two research areas as country-specific. The third level involves efforts to define new knowledge of an applied type that will be useful in a variety of countries. The search for a malaria vaccine would be an example. The fourth level is basic research not oriented toward particular health issues but toward advancing the general stock of knowledge in fields relevant to health.

One can disagree about the particular breakdown. But the notion that people in a country need to be able to identify and deal with their health problems based on the general stock of knowledge is what the Commission meant by country specific health research. The Commission report also discusses the contribution that many developing countries are making toward the broader knowledge base that is useful not only in their own country but more broadly. Scientists at the Oswaldo Cruz Institute in Brazil, led by Carlos Morel, are working at the international research frontier. So are many other third world researchers. This capacity will steadily grow. A national health research program should include both country specific elements and contributions by scientists in developing countries toward the research advances needed by other developing countries and by industrialized countries.

Warren: Let me comment on that. One thing that struck me very much is that interventions that we thought could be applied generally have turned out to be location specific. A particular example is BCG, which was originally thought of as a way to prevent people from getting tuberculosis. A series of trials in different parts of the world have shown that it does not protect against pulmonary tuberculosis in many areas, including southern India.

Bell: It is immensely complex. I don't think we really know what is going on with BCG. It is a very controversial story. It is a perfect example of how little we know about many very important issues.

3

Organizing for Health Rather than Sickness

Donald A. Henderson

Before addressing institutional structure I would like to emphasize how important it is that we recognize the range and complexity of the health problems that confront society. AIDS is an extremely interesting case. We have to bear in mind that AIDS is, if you will, a wild card that has suddenly appeared. In conferences this past year, we have been discussing the potential of other new viruses emerging. Joshua Lederberg raised the question: "Suppose a virus with the character of both the 1918 influenza and the human immune deficiency virus emerged?" It sounds like science fiction. But this is not something that can be ruled out as impossible. Population is another major issue. But our ability to deal with the population problem is closely related to our ability to deal with disease problems of young children. It has not been possible, in any society, to reduce population growth rate without first reducing infant mortality rates.

Our biggest problem in the health area is our capacity to address issues. There is both a lack of suitable forums in which to ask the relevant questions, and limited resources with which to answer the questions. What do I mean? We now have in place a global system of international agricultural research centers. There are only two centers in the field of health that bear in any way a resemblance to the international agricultural research centers. One is the Diarrhoeal Diseases Research Center in Bangladesh, whose budget is around seven or eight million dollars per year. The second is INCAP, the international nutrition center in Guatemala, which has only a vague resemblance to the agricultural research centers. Its budget is only around two to three million dollars. Beyond that, we have individual researchers located in a number of different countries. There are also quite a number of university settings where some work is going on. But the number of research staff in developing countries is limited and financial support is exceedingly limited. Given that many of the studies that are done should be done within a local context one can only conclude that we are very poorly organized to meet health research needs.

In the industrialized countries support for research on the health problems of the tropics is also very limited. Support for health stands in sharp contrast to the AID Title XII program in support of agricultural research with its extensive system of collaborative research networks. There is no equivalent to Title XII in the area of health. The capacity for tropical health research in the United States is seriously deficient (NRC/IM; 1987). We have an aging group of experienced people. But there are few new people coming into the field. There is very little support for such research from any source in spite of the growing tropical disease problems. We also have in this country very little in the way of academic education or research on international health care issues. The former colonial countries - United Kingdom, France, Netherlands, Belgium - that did have strong institutes at one time have let them atrophy. The London School of Hygiene and Tropical Medicine is probably the strongest in Europe. But the London School is not large. Its university funded faculty is hardly more than 50.

Industrialized country support is very limited. US/AID has taken a view that in the health area it would undertake its activities primarily on a contractual basis. And so we've had the growth of a great many private sector organizations - the belt-way bandits - working on contract in a variety of areas with very little support from academic institutions. There is a policy not to fund academic institutions. This stands in sharp contrast with agriculture where so many of the AID supported program and research activities are conducted at academic centers where they contribute both to research and institutional development.

The Department of Health and Human Services has exhibited little leadership. When the Office of International Health, a small office under the Surgeon General, was reorganized a few years ago no one in the field even knew the name of the director. It has been a small office made up of generalists with little relevant experience. The Centers for Disease Control has a small program that is very focused and almost entirely AID contracted. The Institute for Allergy and Infectious disease has a small research program - small, fragmented and without a particularly clear set of goals. Thus within our own government we have not developed a very effective program. We badly need, within the government, a staff that can think more seriously about international health problems and policy.

The World Health Organization (WHO) has been helpful in bringing together groups that have asked strategic questions, it has, in certain areas, been useful in developing a research and action agenda. But WHO plays a different role from FAO, for example. WHO has devoted excessive energies to protecting its own turf - insisting that it is the technical health agency of the United Nations and therefore all health expertise ought to be within WHO. This has led to some very difficult problems with UNICEF, for example, which for a long time was prevented from hiring the technical expertise it needed for its programs. Many WHO staff are political appointees. Some regional offices are highly politicized. WHO is an international body that should be involved in asking questions and mobilizing resources. But its record is very checkered. In some areas it has proved to be a fig leaf used to cover the absence of real activity. Its Tropical Disease Research Program is funded only to the extent of 25 million dollars and covers only six of the tropical diseases. This is an exceedingly small amount of money for such a huge array of problems. The bottom line is that within the health structure we probably could not invent a less effective way to bring our expertise to bear on international health care if we tried.

I have painted a discouraging picture. Let me complicate it a little bit further by going on to say that all of this takes place against a background of completely inadequate health services. Let me make my point a little more bluntly. By and large we have developed a sickness recovery system - not a health care system - but a sickness recovery system. We have in this country highly sophisticated hospitals and clinics. But we are ill-equipped to deal with AIDS or teen-age pregnancy or substance abuse. We do not reach out into the community. When we look at the third world we have hospitals and primary health centers that generally include one or two doctors with a small supporting staff of people - practicing physicians perched somewhere out in the not too distant boonies. They render a curative service to the people who appear at the door. It is very difficult to get many of these centers to even vaccinate children - simple preventative procedures are something they are unaccustomed to providing. So what we have in both the United States and the developing world is not a health care system but a sickness recovery system. And this is the culture of medicine. Doctors who have been indoctrinated in this system are making many of the policy decisions today within the WHO and within our own government.

Arguments go on about the advantages of the vertical programs versus the horizontal programs. A common contention is that we should be providing at every center all the services without emphasizing any particular activity (a horizontal program). This has been one of the silliest policy arguments. It is a carryover from the myth of the practicing local physician who provided all health services - both curative and preventive - to whole families and, in turn, to the community. Recently a number of programs have gotten underway - immunization, Vitamin A and a number of others - that are community based. An effort is made to reach out to the entire community to provide services. Such programs offer promise for the future. But they are very new and date back little more than a decade.

If one is providing community health services one must have data for the community. In medicine this is an alien phenomenon. We just simply do not collect data on a systematic continuing basis on disease. We do not have effective surveillance systems for disease. This is true in the United States as well as in the developing countries. It emphasizes our continuing commitment to a sickness recovery system rather than a health system.

I have presented a rather sad tale about a system that is not well organized to address the problem of health in the community. We have a set of institutions nationally and internationally which are at best of marginal value when valued as providers of health rather than in terms of sickness recovery. The medical community sees its role relative to health quite differently from the way the agricultural community views its role in sustaining and enhancing agricultural production. As we look to the future and the set of issues facing health and agriculture there is a lot here that could be developed by working together. There is a great deal for us to learn. The agricultural extension system is a remarkable institution. We have nothing remotely comparable in the health field. Schools of veterinary medicine are doing a lot of very interesting work on disease prevention. But as far as the schools of medicine are concerned they might as well be on different planets if not on different solar systems. There is almost no effective communication - to the detriment of health. A shift in our focus from the sick patient to the health of the community will be needed if we are to deal effectively with health issues in the future.

Warren: When you mentioned the dichotomy between medical and veterinary schools the question occurred - what about the dichotomy between medical schools and schools of public health?

Henderson: They are at least on different planets.

Sanders: You mentioned a couple international health centers. What do you think of the work of the Onchocerciasis Control Program that works on river blindness in West Africa as a model?

Henderson: That is a program designed to solve a problem. And I think its record of productivity has been quite good. It's a very-high cost, targeted program.

Warren: One of the interesting things about that program was it was started by MacNamara when he was president of the World Bank. During a visit to West Africa he became appalled by river blindness. Years later at a meeting at the Ford Foundation I remember him commenting that, "If there is anything that I think we did right at the World Bank when I was president it was to initiate the Onchocerciasis Control Program." Several of us at the table groaned. We said, "Look the whole system is based on controlling the black fly. Even if you controlled it completely in a local area, they'll come in from elsewhere if you stop." A major omission of the program is that it did not support research on other means of controlling onchocerciasis, such as drugs and vaccines. I am, nevertheless, deeply impressed with what it has achieved by vector control alone, but this will be greatly enhanced by the use of Merck's new antifilarial drug, ivermectin.

Henderson: I want to elaborate further on a common assumption that is made about programs in the health area. Too often it is assumed that we know what to do! It is simply a matter of going out and getting the job done. One of the more successful programs of recent years has been the immunization program. In a period of a little more than 10 years it has moved to a point where children around the world are receiving antigens against six diseases. It has gone from 5% coverage to 70% coverage. But from the beginning, there has been no research component to the program. Except for improving methods for refrigeration there has been no research. Nothing has been done about the vaccines - none are fully satisfactory. The vaccines used in the program are essentially the same vaccines that were available 25 years ago. It's only in the last few years that additional work - very limited - has begun to improve the vaccines. This is a recurrent theme in health. Malaria is a case in point. When that program began it was decreed that no research was necessary. Research was stopped in most of the major centers because program Directors asserted they knew what to do. It was simply an administrative matter of going out and doing it. When I look to the agricultural model, I see many programs for developing new varieties of plants. And there is money invested continually in genetic enhancement. Our record in the health field is to say again and again we know what to do - let's apply it. We keep insisting that we don't need research!

Bell: I would like to ask the agriculturists about the West Africa experience because we've heard the comments on it and certainly in the short run, it has been rather dramatically effective in reducing river blindness. The response, if I remember correctly, was that it initially created a great deal of agricultural development. But there was a report three or four years ago that was considerably less optimistic. What happened?

Sanders: There is a report from the World Bank showing extremely rapid agricultural development in Burkina-Faso in the 1980s in spite of the drought in the early 1980s. And a lot of that is due to French research and marketing investments and to improved methods of fertilization that have been used on cereals. But the several sources of increase in production here have not been clearly separated out by economists. In Burkina-Faso the zone just above the southwest has been a boom area. It is really exciting to be there! There is very rapid development. It may be unstable if the black flies can move back. But the changes in the 1980s have been very dramatic.

Bell: On what scale? Are you talking about a few thousand farms or something large enough to have a regional or national impact? How much affect has it had on agricultural production in West Africa-- 2% or 50%?

Sanders: There was agriculture in these areas before the river blindness outbreak. There has been some exaggeration of the effects of river blindness. But it has not yet been possible to determine how much of the growth in production was due to the resettling of the river valleys, how much to use of fertilizer, and how much to research leading to new practices.

Herdt: I'd like to make a couple points to the health people. The theme that we know what to do and all we have to do is go out and do it has been a very familiar theme in agriculture. Some well known international institutions were set up on the assumption that we know, for example, what to do with rice in West Africa because we have already done it in Asia. Since we know what to do with livestock in Latin America, we can approach the problem in Africa from a technology transfer perspective. Obviously there is some relevant knowledge. Going back to Bell's criticism of Ruttan's agenda--I wonder whether the difference is that we view health as a consumption good and view agriculture as an investment good. Ruttan and several colleagues have been pushing the idea of agricultural research as an investment good for several decades. From the point of view of state and federal support for agricultural research it is an investment. Better health is too often viewed as merely enhancing consumption. I realize that Mark Rosenzweig and the "human capital" school in economics view health research and health improvement as an investment. But there have been very few studies that document the economic benefits of health research. David, this goes back to your opening comment. Ruttan phrased the issue as an investment question. You objected. But health research and health programs can productively be viewed as investments in enhancing the quality of human capital.

Bell: I evidently misled you and perhaps others. My point was that health does not just have investment characteristics. The Health Research Commission strongly argues the point at the beginning of its report. It argues that the significance of health to development has been greatly underplayed. Health is an investment for development. This position is very strongly asserted by the Commission. The health of adults does affect both the number of days they can work and their productivity. One of the most significant aspects of health as an investment is, of course, in children. In the health community I usually speak as an economist. I am impressed that the community has repeatedly observed tremendous impact on health improvements on growth and development in children - what they can learn in school and what they are physically able to do. But it is not well documented. The point is not that health as an investment is not important. The point I was trying to make is that these relationships are much more complex than simply asserting that we are interested in health constraints in agricultural development.

Warren: The Commission started with a bias in favor of health. I share that bias. Neither of the reports of the two great international commissions on development of the last 15 years gave any significant attention to health. In the Pearson Report there was about one page devoted to health; in the Brandt Report, virtually nothing. The one point where health was mentioned in the former was negative, because it stated that improving health merely exacerbates the population problem.

Rosenzweig: I think that is an old perception and those are old reports. The 1990 development report by the World Bank was largely devoted to human resource issues. About half of it was concerned with health. Even so it doesn't have much to say! The problem is that we don't have much of a database. But economists are beginning to look at health as instrumental in terms of its contribution to development.

The basic problem was illustrated in the discussion about the effect of onchocerciasis control - we don't have very good data. And we don't have very much research as a consequence. These are complex interrelations that David talks about. There ought to be out there in the world the ability to measure the returns to these investments. But when you look at all the studies they are almost empty of anything one might regard as hard evidence. I'll talk about this issue in my chapter. But there is a growing perception that it is an important area - a hot topic. The World Bank has come on board and is now sponsoring important data collection efforts - primarily in Africa. Some attention, but not enough, has been paid to the nutritional status of children, using height and weight as indicators, and to getting some rough information of illness and lost days of work for adult populations. So I think our capacity to understand some of these impacts is going to be greatly enhanced in the future.

Ruttan: Your comments represent a useful illustration of the point that it takes about 20 years from the time a fundamental breakthrough in conceptualization occurs before it becomes a leading topic in research agendas.

PART THREE

Disease in the Tropics

4

Agriculture and Health

David Bradley

Bradley: I was intrigued by the comments about the strength of the institutional infrastructure to address research related to crop production in the tropics. But I also recall a conversation by someone who had served on the governing body of the Consultative Group on International Agricultural Research (CGIAR). The comment was that the work at the Centres was rather static but the research networks that focused on more location-specific research were very effective. But perceptions probably change much more rapidly than reality.

I work at the Ross Institute, which was founded in the 1920s, because it was believed "We shall never get all of these indigenous people to work productively on our plantations and make us money unless they are healthy." This was an extremely cold-blooded approach to the fact that you couldn't grow tea cheaply without malaria control. Earlier on, the ports that had just been built on the Malaysian coast were losing money and were liable to close because the mortality rate from malaria among port workers had just reached 800 per thousand. It didn't, at that time, seem necessary to do academic research to prove that disease hindered agricultural production.

Let me draw on another personal illustration. My first job was to find out in Tanzania whether or not urinary schistosomiasis made people ill. Very many people were infected but there was doubt as to whether they were actually unwell. If one looks at the reason why that piece of research was taken up, it was because at the time, everyone said: "Intestinal schistosomiasis is a terrible thing. There is no doubt about that. But no one seems to think that urinary schistosomiasis matters." We did in fact find some interesting and some rather frightening things from the research. About ten years later, other people said some research must be done on intestinal schistosomiasis to find out whether it is making people ill. "We know that urinary schistosomiasis is dreadful but we are not sure about intestinal schistosomiasis." The point of my comment is that without doing the research one can not assess the potential significance of the health problem.

Water resource development and health

My perception of this meeting was somewhat different than that of Dave Bell. I visualized a line that had health constraints at one end and agricultural development at the other with arrows going both ways. The interaction is very complex. If you build a reservoir and irrigate to get more production of rice or some other crop, it produces complex changes in health. The people who live near the place where the dam was built invariably come off worse in every possible way. Even if they end up with higher income, they still seem to come off worse! There are very subtle and complex interactions that I will illustrate in a moment. However, public health interventions are a very blunt instrument. Subtle and complex programs are very, very difficult to design and implement. The social patterns in agricultural production - the way in which people are organized - often have very complex effects on health. You can not reasonably ask an engineer to spend his time on health education, or even to spend much time thinking about health problems because he can't and won't. Any yet one has got to have health considerations incorporated into the way in which agricultural development occurs; otherwise you may be letting yourself in for some very unpleasant shocks. I can illustrate this with malaria.

When the Volta dam was going to be built in Ghana, health problems were anticipated. The Kaiser engineering people brought some experienced health engineers over from the TVA. They recommended a procedure that has worked very well in the United States - dropping the level of the reservoir every week or so to strand the mosquito larvae. The stranded larvae die. The procedure did work very well in the Tennessee Valley. But if it had been adopted in Ghana, it would have enormously increased the population of malaria mosquitos in the Volta dam area, where the mosquitos are puddle breeders. Fortunately, someone pointed out it was going to be too big a lake to vary the levels so the procedure was not adopted. But that isn't the end of the story. There was already an enormous malaria problem. If you increase the number of mosquitos everyone would have gotten bit a few more times. But there would be no increase in the amount of malaria. Because there was already a thousand more malaria transmissions than needed to maintain effective transmission. This is very important if you are going to attempt to reduce malaria by reducing the mosquito population. There are some places where a 95% reduction won't make the slightest bit of difference in the amount of malaria in the human population. Sophisticated advice is needed on such issues. And there is a real problem of matching up sophisticated advice at the village level in a poor country.

The difficult part about most health aspects of agricultural development is that it requires action by other sectors, in particular by engineers and agronomists. It has been possible to make considerable progress over the last 20 years with the engineers. This has partly been because the health workers have made a specific attempt to translate the issues into the language of engineers and also that we have gone on pushing very hard for a long period of time and made a point of going to the meetings at which engineers congregate, rather than talking to the societies concerned with health. The result is that now most engineers concerned with water resource development are quite familiar with the health problems and will at any rate make some sort of effort to respond to advice on possible adverse health impacts of their work. There is still, however, a tendency to pay lip service to health issues by suggesting that a health expert should look at them, whereas what is necessary is that the other disciplines should be sufficiently involved to take appropriate action. By contrast with the engineers, the agronomists still do not want to know about, for example, the health implications of irrigated rice production. These include the effects of rice cultivation on malaria and on Japanese encephalitis. Malaria is transmitted by mosquitoes, some of which breed in rice fields and the main vectors of Japanese encephalitis are typically ricefield breeders.

So the issue of how you deal with transmission due to ricefield mosquitos is very important. The medical entomologists wanted to drain out the ricefields during the growth of the crop at intervals to kill the mosquito larvae. The agronomists said absolutely no way. This would reduce yield very substantially and the only issue that they wanted to think about was rice yield. The effect on production of modifying the water regime would be too great. They just didn't want to face the health issues. By contrast, this has been addressed in China where they do dry out the field during one stage in the production of the crop. The evidence there suggested that this was not only effective in controlling mosquitos but it also has favorable agronomic consequences. The reason for this is that in China they use human excreta as fertilizer. Drying out the field is important to the proper growth of rice when human excreta are used as manure but not when chemical fertilizers are used. This story illustrates two things. First that the interactions between agronomic practice and vector breeding may be extremely complicated, and second that the assumptions which are commonplace among one discipline (such as that the type of manure affects the water regime suitable for growing rice) may be completely unknown to another discipline that is involved in the problem. It takes time to transmit knowledge across disciplines and it also requires people to divert attention to doing this from simply following their own research interests. It has, after all, taken about a generation to get the engineers to deal with health in relation to water resource development.

A third point about health implications of water resource developments for agriculture concerns the scale of activity. Dams are still being built without worrying about health issues, unless you have to borrow from an international organization such as the World Bank that insists on having proper environmental impact assessments carried out. This means that health issues are taken into consideration quite well for very large dams where international borrowing of funds is essential. This doesn't get us too far because for most countries, one large dam is about all they can afford and therefore research on how to cope better has no consumer because there is no second dam into which the conclusions carried over from the first can be fed. By contrast, small dams are usually built without international borrowing and they take place on a very large scale. There were something of the order of 5,000 quite small dams built during the last decade in northern Nigeria. They are very important for local agricultural production but there is no mosquito control, no health assessment and no planning in relation to health at all. The health implications of small-scale developments are a very important issue for research and for the future. Moreover, any research done on small dams has an immediate consumer because many more will be built in the same locality.

A further issue concerns the length of time it takes for the full health implications of a large water resource project to emerge. For example, the Volta Lake in Ghana has taken some 15 years or so to stabilize. Initially, there were transient periods when there were enormous crops of fish that in turn resulted from the insects feeding on the rotting trees that had been immersed by the rising water of the reservoir. These complex systems led to a great deal of eutrophication of the water. This led to large quantities of water-weed, enormously increased snail populations and lots of schistosomiasis transmission combined with very large numbers of immigrant fishermen, both transmitting and catching schistosomiasis. After over a decade, the trees had all rotted and there are now fewer fish, fewer fishermen, less water-weed, and probably less schistosomiasis.

A final issue in relation to agricultural development, particularly through water resource development, is that the traditional approach to health has been through environmental impact assessment. In other words, the interest of the World Bank in schistosomiasis was essentially from an environmental perspective, and to ensure that the environmental change did not lead to a worsening of people's health. A much more constructive approach would be to look at the opportunities for improving health when a water resource development is constructed. The aim would be to add a health opportunity assessment to the traditional environmental impact assessment. This requires thinking of health as a goal of development rather than ill health as a side effect of development that is best avoided.

A typology of tropical disease

Now let me turn more specifically to the tropical diseases. There are two main categories. First there are the diseases of poverty - the sort of disease you would have found in a poor part of London or New York a hundred years ago. These diseases predominate in many parts of the third world today. Often the diseases of poverty are reduced by simply increasing the disposable income of people. Things that they then will do on their own will tend to improve their health. Second, there are the specifically tropical diseases. These are diseases which have a stage in their life cycle outside the human body and which is temperature-dependent. They are usually transmitted by insect or snail vectors, or they may sit in the soil waiting until certain temperature requirements for development are fulfilled. This second group require rather specific action which is often closely related to changes in the environmental associated with agricultural production, and they may also produce chronic infections that can interfere with agricultural production.

Among the tropical diseases, it may be worth distinguishing between the helminthic or worm diseases and everything else that is infectious. In the case of the virus and bacterial and protozoal diseases, you may not be able to tell after a few days of infection whether someone was infected by one virus particle or by several million. By contrast, the worm diseases are quantitative. If you have been exposed to 23 hookworm larvae, you can get up to 23 hookworms but not more unless you are re-exposed, and these helminth infections are quantitative in their effects on health. If you catch a few of them by going out without your shoes on hookworm-infected soil once or twice and not every day, then they are going to be much less important. If, however, you are infected frequently and get several thousand hookworms you are very likely to get seriously ill or even die. It is not just have you got a disease like schistosomiasis, but how many schistosome worms have got you! The same is true of hookworm, river blindness and other worm infections. Getting rid of the last worm is certainly relevant to stopping transmission, but it may not be relevant to sickness or illness.

There has been a great deal of effort put into focusing on the diseases of childhood in the last decade and rightly so, but if we wish to disaggregate mortality and look at the changes in health in terms of their relevance to agricultural productivity, one perhaps needs to look more strongly at the probability of dying between the ages of 15 and 60 rather than just looking at life expectancy at birth.

There are four broad categories of interventions which can affect tropical diseases. These include pesticides to kill the vectors of tropical diseases, chemotherapy to kill the infectious agents, vaccines, and environmental management. They each have different characteristics. Pesticides were the main approach to control in the past. Their effect on transmission was immediate. Whether it has an effect on morbidity or anything else depends on whether you are dealing with a long-lived or a short-lived parasite. If you've got a massive malaria epidemic, as they had in Khartoum at one stage, one of the things you can do is to move in, kill the mosquitoes and stop transmission. This is all right as a rare activity but there have been two problems with insecticides and other pesticides. The first has been the emergence of resistance. The second has been the sheer expense of the insecticide route. In the case of long-lived parasites, for example, there is a need to maintain vector control without any errors for long periods of time.

Chemotherapy has a different pattern. I am sure Ken Warren will speak enthusiastically on the role of chemotherapy in controlling tropical diseases. In the case of schistosomiasis, it has wholly transformed the situation. If one looks at a longer time-scale, the history of non-helminthic diseases suggests that chemotherapy has been effective only in the short term. It has been much less effective in the long term. The great yaws campaigns of the 1950s were very effective. Everyone said yaws has been eradicated. Then a few years ago, suddenly people woke up to the fact that they were getting 5,000 new cases of yaws each month in Ghana. The disease had vanished politically a long time before it vanished biologically. One of the problems indeed with malaria control is that unless people are dying around you, it is very difficult to maintain a high budget. Thus chemotherapy is in some ways the most problematic of the attempts at control because it has turned out in so many cases to be rather a short-term solution to problems. Certainly it should be more successful in worms which build up resistance to chemotherapeutic agents more slowly than do bacteria and protozoa, and maybe chemotherapy will prove more successful in the future. Certainly it is the way to deal with morbidity and the reduction of mortality rapidly.

Vaccines are still the great hope. For the bacterial and virus diseases, they are indeed extremely effective. In the case of the human parasitic diseases, they remain a hope for the future. We were going to have a malaria vaccine before the end of the last decade and we still don't have it. We can be certain that we won't have one in operational use for another few years but I remain optimistic that in time we shall find an effective one. Vaccines are moving targets. We know empirically already that there are good effective vaccines against many organisms and indeed, in the case of bacterial and viral vaccines, improving the ones we have already is a large challenge. Surprisingly, organisms so far have not produced a lot of resistance against vaccines. There is no major parallel with the emergence of insecticide or drug resistance though that is not to say that it cannot occur.

Environmental management is the last type of control I want to discuss. Because it requires a fair amount of behavioral change it often requires a high amount of discipline. Requiring a positive discipline rather than taking a pill or having someone spray your house has a very checkered history. Yet it seems to me the approach that, over the longer run, has gotten rid of diseases more permanently - particularly from the presently developed countries.

Finally, I would just like to pick up the issue of assessment of effects. One of the reasons why the health people have not done thorough evaluations of effect of health on productivity of a comprehensive nature is because they are very expensive. If you are going to spend 20 million dollars most of us are reluctant to spent much of it on evaluation. Is that the best way to spend part of the 20 million dollars? It would be interesting to know from the agriculturists how they would respond. Unless one is going to spend enormous sums it will be necessary to be very selective. There are certain situations where evaluations can be done at relatively low cost. If you are really interested in knowing whether a child's illness affects productivity, it might be best go to a tea garden where ladies who prepare the tea have children and the amount of tea a woman plucks is weighed accurately. Her pay depends on it. In this situation, you could very readily measure the effect of hookworm on a child's productivity.

Ruttan: Let me make one skeptical comment. We have put quite a bit of effort, beginning about 25 years ago, into studies of rates of return to agricultural research. And then some young examiner at the Office of Management and Budget said to a group of us as we were presenting these results, "It is not the job of the government to make a profit!" Budget constraints had replaced economic benefits as a criterion for policy decisions!

Warren: David, I was interested in your comment about the engineers understanding of health issues a decade or so ago. What was it that brought that issue to their attention?

Bradley: One was the steady pushing of the issues. A few engineers got interested whenever there was a meeting about dam building and some health person was on the program. We had a great deal of trouble rephrasing tropical medicine issues in terms of engineering language. We concocted a classification which wasn't in terms of biological classification but categories of intervention - what you have to do to stop them. The principle is to translate into the language of the recipient. But I think that what really produced the change was that people who wrote terms of reference for engineers started writing in health considerations. It became part of the environmental statements. People who write terms of reference are the ultimate people who change the way in which things are perceived.

Warren: The involvement of engineers goes back to the turn of the century when we started using chlorine and sand filtration to reduce the number of infectious organisms responsible for many of the diarrheal diseases. The engineer at that time who was involved in the activities was called a public health engineer (in the United States).

The name later was changed to sanitary engineer and more recently it became environmental engineer. But he is still basically a public health engineer.

Bradley: That's a very good point. There has been a sequence of three groups of engineers - there are the public health engineers who have continued along the whole time but they don't get involved in building dams. There are the malaria engineers, who were largely located in the United States, who are concerned with controlling malaria. They died out almost completely after 1945 when DDT was found. We are having to reinvent them. Then there are the civil engineers who build dams and other civil works. When I was speaking about getting them educated I was only speaking about the civil engineers who build dams.

Straub: Twenty or thirty years ago Professor Fere at Harvard was responsible for setting up some public health training activities for engineers in England and other European countries. The development occurred here first and then went over to Europe.

It is useful to distinguish between the big engineers and the little engineers. The big engineers do projects that are enormously expensive. It is not always easy to consider health in the equation because these projects are so massive. The Gazira project in the Sudan is an example. When they started they were very concerned about schistosomiasis. They did everything they could, engineering wise, and in other ways, to try and prevent schistosomiasis and they failed.

Sanders: Why were they so unsuccessful in controlling schistosomiasis? Was it lack of knowledge or were there too many interacting factors?

Straub: They didn't know how. For instance, they tried to prevent snails from getting into the canal by building barriers. But that didn't work. Snails can hop over on the feet of birds. The specific engineering measures that they took didn't work.

Bradley: There are two very large schemes in the Sudan. The first one was not designed with schistosomiasis in mind, but the second one was. There were certainly some naive aspects of it, but for whatever reason, the level of schistosomiasis is very much lower in the second scheme. So an effort was made that did seem to produce some effect.

Straub: One thing you can do in going about solving the problem is to install concrete canal linings. But that is very expensive.

Bradley: In my earlier comments I didn't adequately address the question that had been put to me "What are the effects of tropical disease on agriculture?" There is a group of diseases which seem to directly restrict agricultural activity. I think there are only three. One is onchocerciasis, the second is trypanosomiasis, and, finally there is AIDS. Then there's a group that affect a sufficiently large part of a population that they might reduce productivity overall. One is hookworm. Another is malaria. The classic example is in the Gambia where people are extremely hungry just before planting season. Labor requirements are enormously high. It is also the peak malarial season. We haven't really addressed the issues of cyclical disaster situations.

There is a second group that affect mortality during the working years of life. Some of the key ones are (a) tuberculosis, (b) AIDS in the areas where it is being heterosexually transmitted, (c) possibly encephalitis, and (d) drug-resistant malaria is becoming an increasing problem at the moment. There are lots more things that could be added to that. To put numbers on them, one would have to do quite a bit more work.

Finally there are other things that cause disability to individuals - things like polio and trauma. There are things like malaria and diarrhea - the things which cause loss of working time and require time to care for other people.

Let me now turn to the effect of agriculture. The question of where periodic illness fits in was a question in my mind. There are problems which are potentially made worse by agricultural change, and therefore need some prophylactic action - schistomalaria, trauma, insecticide toxicity and excretory use. I think it's helpful to distinguish between two possible adverse effects. One is where you've got specific diseases getting worse due to a particular way of developing agriculture, and the other is by effects on the disposable income of certain parts of the community. If the rich get richer and the poor get poorer it is always a bad thing for health.

The third issue is your question of what are the areas that we should be thinking about for the next generation, particularly the areas for agricultural research concern. What about farming systems that are appropriate for refugees and camp-dwellers? The number of people who are living in very confined areas in pretty awful circumstances--not the sort of thing that any agricultural research institute would like, but nevertheless, maybe there's something that can be said, maybe there's nothing that can be said on that. Similarly, the problem of adapting your farming systems or developing farming systems which don't make the very poor poorer.

5

Vaccines, Drugs and Tropical Disease

Kenneth S. Warren

Warren: Historically, tropical disease research was almost wholly concerned with parasitic diseases. Parasitic diseases are those caused by protozoa and helminths. The schools of tropical medicine in the U.S. and Europe focused largely on those organisms. The interesting thing is that when we began to look more carefully at mortality figures we found out that the greatest killers in the developing world were not the parasitic diseases, but the diarrheal diseases and acute respiratory infections caused largely by bacteria and viruses.

There are about 32 million deaths a year from all infectious diseases in the developing world. Of those, the only major parasitic infectious disease is malaria - which causes about 1.2 million deaths. The parasitic diseases, however, caused an enormous amount of morbidity due to their prevalence and chronicity. Individuals may not only have one of them, but, in many cases, seven or eight different parasites. The data on morbidity is a scandal in public health, as we have failed to develop methods of measuring it accurately. If we can't measure morbidity, we can't convince anybody that these diseases are worth spending any money on.

Now that gets us to economics. I would like to suggest to the economists that medical and health research is relatively cheap! Furthermore, research offers enormous hope in being able to develop new methods, and improved and cheaper methods for dealing with these problems. So I think that we should not hesitate to insist that a small but significant proportion of the money going into any control effort should be devoted to research.

We're still dealing with vaccines that are twenty to fifty years old, and that are difficult to administer. If we had better vaccines, we could work much more efficiently and cheaply. I'm taking a more positive view on this than anybody has expressed here so far. I think that we can do many things that are positive and relatively inexpensive and can improve health drastically in the developing world.

We've had about six vaccines that are highly efficacious. After the great success of the smallpox eradication campaign, WHO started the Expanded Programme on Immunization. A problem was that the immunization campaigns, though planned very carefully, were moving too slowly. UNICEF's child survival program, has, in contrast, been particularly successful. Through The Task Force for Child Survival which coordinated the efforts of WHO, UNICEF, UNDP, the World Bank, the Rockefeller Foundation, and Rotary International, the immunization rate rose from 20% in 1984 to more than 70% five years later. It should reach 80% this year. These campaigns are directed toward mortality, although they do affect morbidity - particularly with respect to polio.

The vaccines that we have available now work, but if we could improve them we could drastically increase coverage and lower costs. We now immunize children against measles at about 9 months of age. But severe measles occurs in younger infants. There is a new vaccine called Edmonston-Zagreb that can be used to immunize children at 4 or 5 months. It is also necessary to reduce the number of doses of most of the vaccines and to eliminate the toxicity and side effects of the pertussis and polio vaccines.

The first genetically engineered vaccine may be the next vaccine added to the Expanded Programme on Immunization. Hepatitis can be transmitted directly from the mother to the infant, but death doesn't occur at that point - it occurs thirty or forty years later when the patients die of cancer of the liver. Liver carcinoma is one of the most prevalent forms of cancer in China, Southeast Asia, and Africa.

The important thing is that the immunization campaigns have been shown to work. Annual or semi-annual campaigns have been developed and the infrastructure now exists to maintain these high levels of immunization. Now that the implementation systems are in place, the next priority is new vaccines. The two areas that we really have to do more on are diarrheal diseases and acute respiratory infections. The one vaccine in the diarrheal area that looks somewhat hopeful is that for the rotaviruses which cause about 15-20 percent of lethal diarrheal episodes among infants and young children. Other necessary vaccines are those for enterotoxigenic E. coli and cholera.

For acute respiratory infections, the age of the infant is a problem. The polysaccharide vaccines for pneumococcal pneumonia cannot be administered to children under two years of age. A new polysaccharide vaccine for H influenza B has been hitched to a protein, and can now be given at a younger age. This might be done for the pneumococcal vaccine as well. Work is also underway on two of the major viral infections, parainfluenza and respiratory syncytial virus.

In the meanwhile, we have developed a variety of drugs with which we can control virtually all of the helminthic (worm) infections of mankind in a very simple and relatively cheap way. We now have three broad spectrum anthelmintic drugs which effect many different parasites simultaneously. And they overlap so that most of the helminths infections can be treated using these three different drugs.

The other interesting thing about these drugs is that they're all oral, and are all single dose and non-toxic. David Bradley emphasized that if you have just a few worms, you don't get sick. With all of the worm infections, it is the small proportion of individuals with heavy infections that become ill. The distribution of worms in populations is not a bell shaped curve. It's a curve in which only about 10% of people at the end have heavy infections. Most of the 10% are school age children. The strategy used by the Rockefeller Foundation in the 1920s to control hookworm in the southern United States focused on that 10 percent. Now you can give a single drug, albendazole, to control the major geohelminths, hookworms, ascaris and trichuris, simultaneously. Praziquantel can be used for the control of schistosomiasis, and in most cases, requires one treatment at three to four years. The drug is also effective against liver flukes (except Fasciola), lung flukes and intestinal flukes, and all of the cestodes (tapeworms). Ivermectin will control onchocerciasis and the filariases, as well as several of the geohelminths. It would have to be given about once a year or perhaps every 2 years. It is hoped that anthelmintic campaigns will be developed through the school systems. They will keep helminths well below the level where morbidity occurs. An interesting thing about the drug campaigns is that if you are giving the pills frequently enough, you can also include some important micronutrients such as Vitamin A and iodine. Furthermore, these campaigns would be relatively inexpensive.

At this point, it is worth mentioning that it is possible that the next disease that will be eradicated is a tropical parasitic disease called dracunculiasis, or Guinea worm, or "the fiery serpent"! It is a disease of agricultural areas. Its method of transmission makes relatively

40

simple methods of control viable. Finally, I would like to say that genetic engineering gives us a whole new set of powerful methods for disease control. This field only began 15 years ago. It is amazing how much we've learned about AIDS in the short time it's been around. Think where we would be if AIDS had been discovered 20 years ago, before the techniques of molecular biology were available. Molecular biology not only offers remarkable opportunities to produce vaccines, but also key methods for producing new drugs. Most drugs act through protein receptors and enzymes. Molecular biology will give us not only the primary structure of these strings of amino acids, but also the tertiary three dimensional structure. Once we know the shape of the binding sites of the enzymes and receptors, we can design drugs specifically for those shapes. In the meanwhile, molecular biology is developing new techniques, such as the polymerase chain reaction, with which a few molecules can be multiplied a billion fold within a short time.

I feel that with the implementation systems that can be developed for the delivery of drugs and vaccines, with more effective use of what we have now, and with the addition of what molecular biology and other new techniques can bring us, the future looks bright for the significant control of large numbers of specific tropical diseases.

Allen: Do you have cost figures on the anti-helminth drugs?

Warren: Yes. If you tried to purchase Praziquantel in pharmacies in the developing world, it would be as much as 10 dollars a dose. Because WHO did the major testing of this drug, it is available for mass chemotherapy at a much lower price. The price is declining further because Korea is now producing it in large amounts. Ivermectin is now being given free for mass treatment of onchocerciasis in West Africa. Albendazole is fairly expensive, but there are similar drugs that work almost as well. The basic point is that if we could establish this strategy of distributing these drugs widely - we are talking of treating about one billion children a year - the prices should go down to a few cents a dose. When vaccine purchasing systems were established by UNICEF, the cost of vaccines went down drastically to the present level of 36 cents for the EPI six.

Bradley: If I were Gene Allen, I would find that a rather unsatisfactory answer. Is there some place where the current state of testing these things and the current cost estimates are summarized? It is a complicated story.

Warren: The World Bank is publishing a series of studies of the cost effectiveness of control of the major diseases of the developing world. There will be a paper on helminths which provides many of the details.

Bradley: The point that I was making is that, historically, chemotherapy has been relatively successful in the short run, and relatively unsuccessful in the longer run. But I do agree that helminths are likely to acquire resistance to drugs much more slowly than protozoa and bacteria. I also have some concerns about their feasibility, on a long term basis, from a distributional perspective. Efforts to maintain antimalaria drug treatment on a regular basis for periods of time have been rather discouraging. The first trials usually get very high compliance rates but they tend to fade off. Once you get to the stage where it becomes routine and bureaucratic, then it becomes less effective. I certainly feel if you are going to have regular distribution through the school system, and not the health care system, it will depend on having universal primary education. But I strongly support the view that is the way to go on rather than with the health care system.

Henderson: The question of how to sustain a large scale program is crucial. The belief, at the present time, is that we need a common health care system that provides the drugs through conventional health care providers - mainly the private doctor. The fact that it doesn't work is not questioned! There is now an eight year experiment in Brazil where two days each year they have a program to vaccinate all children throughout the country. They are still reaching about 90% of the children. The question we have to face is what imaginative alternative approaches might be applied to the problem, rather than to ask how to reform the existing system. There are efforts now being made to test alternative methods for delivery of other health care interventions. The anti-helminth campaign is an interesting and innovative way to deal with parasitic disease problems. Not to eradicate or cure, but only to prevent serious illness. There are those who see this as being detrimental to the primary health care system because, if you will, it will be providing drugs outside of our conventional "guild of physicians" system.

Existing vaccines are cheap, but they are all vaccines that were developed for use in industrialized countries. We are paying only the incremental costs for their production now. What we are concerned about is what happens with a malaria vaccine, for example, which has little application in the industrialized world and whose costs, therefore, will need to be borne fully by developing countries. I think we are a bit worried with what may happen with an AIDS vaccine and how that might be paid for because we are by no means certain of how extensively that vaccine will be used in the industrialized countries. The vaccines being discussed at the moment are likely to be expensive. There is no one looking at this problem all the way from the research and development side to the application side.

Bell: In our recent discussion, we have still been talking about delivering health services - delivering vaccines or whatever. We talk about using schools, using the primary health care system and using hospitals. There has been a serious underestimation of the problem of educating people to know if and what they need. In the U.S., measles vaccine hasn't worked because families haven't realized they need it. Consequently, they haven't gotten their kids immunized and measles keeps coming back. We have numerous cases and minor epidemics. The point of all this is that too much thinking in health stops with the medical services. It does not include addressing the question of whether health education can be thought of much more significantly in terms of health motivation or health understanding

The reason I intervene now is that Don Henderson made a reference to the field of agriculture. If you develop an improved variety of rice, there is institutional support for getting it to the farmers, educating them about its use. And they have financial motivations to seek out the knowledge and adopt it. Is there anything similar in the health field? We are inclined to think of families as passive. They get health done to them.

Warren: As you know, Jim Grant at UNICEF is one of the world's great salesmen. When we got into these immunization campaigns, we were all thinking along the traditional lines - which was to put it through the primary health care system. Albert Sabin, however, started polio treatment campaigns in Brazil. Colombia did the same, publicizing the campaigns on radio and television; they actually worked. If we apply these effective drugs for helminths, we will need to develop an education campaign.

Rosenzweig: I think the point about the demand side is very good, but there has been some confusion about the supply side. You shift from the issue of the incentives or motivations for households to use these preventive or curative inputs, to the supply of information as if that was again the principal constraint. It may not be. Let me take the agricultural example. It is true that with improved crop varieties, it is in the farmer's own interest to adopt them. It is also true the most educated seem to adopt first. But for several of the diseases, we have been discussing the returns may not seem very apparent to the households that will have to make the decisions. I think we will have to pay a lot more attention to the returns to these households, given the kind of economic and technological environments that they are in the agricultural example again. The green revolution induced a lot more investment in schooling on the part of households. Households perceived that there was a higher return to schooling than before. It seems to me that the ignorance of the incentives and the technology will tend to lead to mistakes in designing campaigns that concentrate just on pushing particular inputs.

Henderson: Let me follow on that. I think there is a feeling that somehow, if people only knew what was good for them, they would do it. The fact is that we know that most people don't! Around this table here there are a number of us who don't!

Warren: You know that D.A. Henderson is one of the world's great smokers!

Henderson: Ex-smoker! The success of the smallpox campaign was due, in part, to marketing and merchandising. If we got the vaccine into the villages and obtained the support of village leaders, we never failed to get 90% acceptance. It was not a problem. For all of the comments about resistance--it wasn't there at all. If we turn to the United States and the measles vaccine that David was talking about, we find that our coverage for 2 year old kids is about 60%. In San Antonio, where there is an effective information program, it is close to 90%. In some of the cities with poor records, we find that they won't give the vaccine to children unless they are enrolled in a well baby clinic. They are telling the parents that they must get the vaccine, but we're going to keep you from getting it every way we can. They have hours which are convenient to nobody but to the people who are administering the vaccine. How, without making it available conveniently, can people accept it? On the other side, we have many who argue that if the people only understood, they would bring their children. Yet we have figured out every possible way to block them. I support the education. But I don't think that's the real barrier.

Rosenzweig: In a low income environment, the value of children may be low. The value of time is low where people do not place a high value on human life. Rationally or irrationally, it's not clear whether it's in their interest to pay $20 to lower the probability of preventing death or morbidity.

Henderson: That is why the World Bank, in its financing of health services, recommends that preventative services be provided at no cost, whereas curative services should be charged for.

Osterholm: I would like to follow up on a point that Don Henderson made. I think the issue with measles is not just an issue of marketing the vaccine, but it's also a perceived risk. The perceived risk is very high. In the inner cities right now, we usually see less than 40% immunization rates for measles; elsewhere in Minnesota, it's 98%. We just had a big measles outbreak here. It's very hard to get parents to bring a well child in when the only care that they have access to is emergency medical care. In families who are economically unsure, or emotionally troubled, or where there are drug problems, immunization is very far down the road in terms of their priorities. The most immediate problem in many cases may simply be food. It may be a sustained drug habit. That's the real problem. Getting access to care is a major issue. Marketing is very much a secondary issue. Once we can get in there, we can get the vaccine administered. But we have to have access to these children. In the developing world, you talked about Brazil, it may be very different. In this country, we have a whole series of patient rights issues. In Brazil, you may be able to line up 30,000 kids and hit them just like that with no time spent in long lengthy periods to try to explain to parents what the reactions might be. We have a whole different situation here in this country.

Henderson: That's not the way it works in Brazil!

Osterholm: I'm not saying it works in Brazil that way. What I am arguing is that there is a continuum. Why don't the physicians want to give vaccines here in this country? Because they are worried about getting tied into a liability suit. And this has become part of the access to care issue.

Chin: I would like to make a comment about the AIDS vaccine that Don Henderson alluded to. I believe that we won't have to worry about having it in this next decade. There is also an interesting technical facet, because it will be totally different from other types of vaccines and approaches. Besides, the question of whose children you give it to--obviously someone else's children, not your own--there will be an efficacy issue. I submit that whenever a vaccine ever becomes available for AIDS, it's not going to be a 100% effective vaccine. We will be lucky if it's 80% effective. Even if the implementation rate is 90%, with an 80-90% effective vaccine, you will not be able to ease up on efforts to achieve behavior changes. We will need to continue with the prevention package that we're currently engaged in, as well as introducing the vaccine if it ever becomes available. I'm a pessimist about the availability of an AIDS vaccine in my professional lifetime.

6

Where Do We Stand on AIDS?

James Chin

Chin: I serve as Chief of Surveillance Forecasting and Impact Assessment for WHO's Global Programme on AIDS (GPA). I have been a one person unit for the past several years, tracking the global trends of HIV infection and AIDS, and developing surveillance methods to try to come up with estimates of what is the current situation and pattern. What are the short term trends that we can have some reliance upon? What do these numbers imply in terms of morbidity and mortality? What population changes will result from HIV infection and AIDS? The literature on AIDS is simply overwhelming. I can't keep up with it. Much of the literature is not relevant to the public health issues that we are addressing. I let the biomedical research units look at biomedical issues. I am concerned with epidemiology and surveillance--I see very little literature on these topics, especially for developing countries. So what I will do is give you my own personal and professional perspective on what I believe to be the current situation. You can translate that into what impact it will have on agricultural systems.

There are tremendous misconceptions about when and where HIV began to spread. If there is one thing we are sure of it is that HIV and AIDS originated in some other country! There is universal agreement on that. But that's the only thing we can agree upon. From the WHO perspective, it doesn't matter whether it was in a KGB laboratory, or a CIA laboratory, or from Mars, or from San Francisco, or Central Africa. The important thing, at least from the public health perspective, is when did it begin to spread extensively. In the industrialized countries and in Central and Eastern Africa, extensive spreading probably did not begin until the mid-1970s. HIV has probably been present for at least a millennia in Central Africa.

That's not official. That's my personal opinion. There is evidence that the virus was present at low levels in rural parts of Africa in the 1970s. A follow-up in the same population in the mid-1980s showed similar low levels. This suggests that distribution and the prevalence of HIV infections is very dependent on where the virus began to spread in the population and on the sexual and drug behavior in the population. If the virus had not appeared in San Francisco or New York it wouldn't have changed what has been happening in Africa. What would have changed would have been sensitivity to its importance. It was recognized among gay men in San Francisco in the early 1980s. But in retrospect we know from other laboratory data that the virus was already fairly well entrenched in parts of the Caribbean, Central Africa, and some European cities in the late 1970s. So by the time of recognition in 1981, it had already been fairly well seeded in major parts of the world.

We do not have complete information on what is the infection rate or rate of progression of those infected with AIDS. When you conduct studies primarily of people with hemophilia - of gay men in San Francisco or of other gay cohorts - the data suggests that about 50% of individuals infected will develop AIDS within a ten year period. That's about the length of the observation period. We don't know beyond ten years whether the vast majority of those infected will develop AIDS or not. The assumption is that the vast majority will but we just don't know. For projection purposes we make that assumption. For short term projections progression beyond 10 years really doesn't matter because the majority of the HIV infections throughout the world have occurred within the past ten years. The extensive spread started around 1980. The vast majority of infected persons have not been infected for ten years. So we do have at least some measurement of the potential impact. If we can measure how many people are infected and come up with an estimate of the proportion who will ultimately develop AIDS, then we can begin to give a general dimension to the problem. The patterns are quite clear. What we saw in the industrialized countries probably was a very sharp burst of infections among the gay population, and shortly after, among some of the intravenous (IV) drug populations. Whether that has declined in the IV drug population we don't know. But it clearly has declined in terms of new infections in the gay population. But with a mean incubation period of ten years we are still harvesting that tremendous increase in HIV infections in the early-to mid-1980s. It will not be until sometime in the early- to mid-1990s that a major reduction of AIDS cases can occur in the industrialized countries.

One thing I think is clear--we now have grasped the general dimensions of the AIDS problem in industrialized countries. It's not going to wipe out civilization. It will be a major problem, and it will primarily affect the 20-40 year old male population in the cities. The trend in the future will, I think, be totally different. I think it will occur mainly through heterosexual transmission and be concentrated in inner city populations that have high rates of sexually-transmitted diseases. An interesting finding is that if you look at all of the estimates HIV infection rates in industrialized countries have been going down gradually. In the U.K. the 1987 estimate was between 50 and 80 thousand. When the Cox report came out the estimate dropped to 20-50 thousand. Similarly, in the United States the first estimate, made in 1986, suggested that it was between 1.0 and 1.5 million. More recent estimates say that it could be as low as a half million. By modeling, you can generate the current number of AIDS cases in the U.S. with about 500 thousand infections in 1985 or 1986. The proof will be the actual numbers for AIDS in the 1990s in the U.S. Most epidemiologists do not believe that the current prevalence of HIV infections in the U.S.is more than 1.5 million. Therein lies the relative degree of accuracy in serological surveys. But for public health purposes, in terms of general policy and the general impact in the U.S., it sounds ridiculous to say it doesn't matter whether it's a half million or a million in the short run. The numbers of AIDS cases that will be occurring over the next several years will be still substantial. Current programs and policies will have to improve.

The situation in Africa is totally different. It was primarily an urban phenomenon in the early- to mid-1980s. But over the past few years, we are getting very disturbing serological data from rural areas indicating a significant increase. We made a large serological analysis of the data for Africa in 1988 and came up with an estimate of about 2.5 million HIV infections in sub-Saharan Africa. We are in the process now of looking at all of the more recent serological survey data. Our estimates will be finished in the next two or three months, but I can tell you that in the ten most heavily impacted countries in 1987 there were about 1.7 million cases according to our estimate.

Warren: I would regard this precise international data as very inaccurate!

Chin: I can assure you that with regards to officially reported AIDS cases, if I get one number wrong a minister of health will be on the phone saying how come we reported 43 and you show 44? Then I have to do a series of calls to the regional offices to get it rectified. But I don't really believe the data either. But if we believe that there are 4 or 5 million infections in Africa, we can come up with some approximation of how many cases of AIDS should occur in the next three or four years. We are providing the data to the African countries for their policy use. Over the last several years, the pendulum has swung from total denial to recognition that they have significant problems. They now want more support. Part of the problem is that the support that they want and need will be for health care. Let me give you an estimate. We believe that up to now there may be 400,000 AIDS cases. Only 65,000 have been recorded.

Henderson: Are the 4-5 million infections you estimate among adults? If so, what is the total adult population?

Chin: About 200 million adults. I emphasize adults. The whole pediatric problem is a separate estimation and modeling issue. We have made some predictions on what might be the mortality impact. In some central African cities at the present time, the total adult mortality rate is almost double because of AIDS. Some of these cities have an adult mortality rate of something like 5 per thousand per year. We have a few situations in our modeling estimates that mortality due to AIDS alone is close to 5 per thousand per year. In terms of child mortality we believe that by the mid-1990s, child mortality will rise by about 50%.

Bradley: Would that be from AIDS directly or indirectly?

Chin: When we modeled child mortality, we took into account competing sources. With adult mortality we didn't. Child mortality, at least over the next decade, is projected to come down significantly. But when you add in HIV and AIDS, it stays near present levels. The major focus of the problem worldwide will be in Central and Eastern Africa. In these areas, the infection in the rural population has been much less than in the urban. But recent serological surveys show a closing of the gap. So instead of 10% of the urban rate now infection in rural areas may be heading toward 20-30, and in some areas almost 50%, of the urban rate. In a country like Zambia, there is very little urban-rural differential because it is a relatively small country. But even in the most significantly impacted Central African country, we do not believe that, based on these kinds of numbers, there will be a major impact on population growth, at least during this century. If the spread into the rural areas intensifies in this decade then toward the year 2020 it could conceivably result in a negative impact on population growth. FAO has been looking at certain agricultural systems that are labor sensitive. Even in their report, in which they assumed larger numbers than we had estimated, they didn't think during the decade of the 1990s there would be a significant impact. But clearly beyond the year 2000 there could be a significant agricultural labor loss.

Sanders: What about the selective impact on blue and white collar workers in urban areas?

Chin: In contrast to a lot of other diseases, where the higher socioeconomic group is under-represented, they may be over-represented with AIDS. The data are really not all in, but they certainly are not under-represented. When we talk about 20 to 30 percent of the adult population, we're talking about the doctors, lawyers, and technical people. That selected impact will have a tremendous economic impact.

Sanders: Is that being discussed in these countries?

Chin: They are just beginning to get out of the denial phase. We provided them with a lot of numbers in 1989 and we didn't get a single reaction. We will do it again with our new estimates. We are getting a lot of interest and concern now. What we're doing right now is a very detailed estimate for every single country in terms of what we estimate are the number of HIV infections and the number of pediatric cases and short term projections to about the mid-1990s. We do not make any projections as to how high HIV infections will go in the next five years. Maybe two years from now we'll take a look at the serological data again, just to see what we think the level has reached, and then make projections of AIDS cases two or three years beyond that data. We keep trying but we know of no valid way of modeling the increase in HIV infections except to just assume that it's going to continue. But we don't know the shape of the diffusion curve.

Ruttan: Can you say anything about secondary effects? One is the effect, not only on the people who are infected, but on other family members. Another is the effect on competition for health resources.

Chin: I think we're looking at a situation in some of the Central African cities where 20% to 50% of the inpatients may be due to HIV related illness right now. We expect that to go up more as the number of AIDS cases rises in some of these urban areas. We're doing separate modeling on the whole problem of the AIDS related orphans. There are situations where mothers who are infected transmit the infection to a limited proportion of their infants. The mother will likely die within about ten years from AIDS, leaving orphans behind. And those numbers are staggering.

Ruttan: What about the competition for health resources?

Chin: The competition will continue to grow. We are trying to provide projection of HIV related diseases and illnesses to the countries for their planning purposes. Up to 20%-40% of their patients now are HIV related. We make projections of two-fold or three-fold increases in clinical cases over the next several years. Some policy choices will have to be made in terms of how to provide care for these individuals.

Osterholm: There is considerable controversy about how much is spent on AIDS. Public health has traditionally been concerned about both quantity of life and the quality of life. One of the functions of public health is to look at what kills people early in life--what kills them young. We use a concept of potential life loss. It is nothing more than age 65 minus the age at which people die. In Minnesota, with 4.2 million people, half of whom live in the Twin Cities metropolitan area, AIDS is already the number one cause of potential life loss among men between the ages of 20 and 64. It has outdone heart disease in Minneapolis. In St. Paul, it's second, and by next year will rank first among men between the ages of 20 and 65. This is an unbelievable trend. It's unheard of in modern public health history. If you go to the rest of the metropolitan area outside of Minneapolis and St. Paul where the concentration of AIDS is much lower, it barely makes the top five. If you go to greater Minnesota, it doesn't even hit the top 30. So you have this great disproportion in years potential life loss due to AIDS relative to other diseases. In Africa, one of the things that will be interesting, given a shorter life expectancy, would be to try to put AIDS into a years of potential life loss model. With this sexually active population being at risk, which is younger adults, it even makes the disparity worse than implied by mortality numbers. We will be keeping a less productive older cohort alive during the next 10 to 15 years, while losing large numbers of the most productive part of the population. Is this a fair assessment?

Chin: It goes back to looking at the numbers again. For New York City and San Francisco for males and females in the 25-34 year old age group, AIDS is the leading cause of death. It's been that way for the last several years. But from a global perspective it is small in terms of absolute numbers. We have been trying to talk to the Asian countries in the development of their AIDS programs. It fell on deaf ears, except in Thailand and India, where they now are seeing high infection rates in their population. Most of Asia will not experience in this decade large numbers of AIDS cases.

Johnston: Is it correct that the percentage of infected women who transmit the infection to newborns is less than 50%?

Chin: It's a moving target. The data suggests that as the woman develops clinical symptoms, her infectivity increases, so some of the initial cohort studies may have been in populations where a lot of the women were recently infected. Nobody really has the answer to what is the overall transmission rate.

Warren: We have heard about the urban-rural differential in Minnesota. What about Africa?

Chin: What we're finding is that it depends on the country. The British left behind a lot of good things. One of them was roads. So countries like Zimbabwe and Zambia are relatively small countries with good road systems. The urban and rural differential in Zimbabwe and Zambia is not very great. In Zaire, there is a tremendous urban-rural differential. In Uganda, the survey which was completed in 1988 shows a lot of rural areas with 10-12% of sexually active adults with AIDS. It lagged several years behind the urban epidemic.

Bell: We see newspaper accounts, heart-rending stories about particular villages with much higher numbers than in the urban areas.

Do we have the socioeconomic studies of what is happening in places like that? Isn't it a matter of the whole social fabric and the food system rather than just the new numbers?

Chin: The Raikai district has been over-studied. Officials in Uganda now refuse to let anybody in unless they will do something materially to assist. UNICEF is interested in trying to assist with the AIDS related orphan problem. Africa is going to have to respond to this issue by the development of new social institutions that have not been necessary in the past. Supposedly the family and the community took care of each other.

Bell: There really ought to be more African investigators.

Chin: There should be. In the coming years, there will be. But up to now the AIDS problem has been avoided. Now it's being accepted. Unfortunately it's at a time that support for a lot of the programs are being cut back.

Bochniarz: I'd like to come back to your estimates. What is the situation in the socialist countries?

Chin: In Eastern Europe, the situation is totally different. From all of the data that are available--and there are lots of data--the general infection levels of HIV are low in general throughout Eastern Europe. The problem, at least in the USSR and in Romania, is a Murphy's Law situation. You had blood donors who likely were infected out of the country or had sexual contact with infected foreigners. Health workers were treating children with multiple injections of blood, plus multiple uses of needles. We have had spread of HIV like this in a few pockets. But when you look at the whole Eastern Europe medical care system, it wasn't bad practice. We believe that perhaps there could be a significant round of hepatitis B transmissions over the next several years. It's an interesting thing to look at and, like the hemophilia situation, very unfortunate. But I think that, in general, Eastern Europe will experience low HIV infection rates. We will have to continue to monitor it. The USSR has tested 50-80 million people, mostly pregnant women, and found virtually no infections.

Bradley: I want to come back to your question on secondary effects. One problem for the orphans is not only are they orphans but also they are stigmatized because of what their parents have died from. Whether HIV positive or negative they are not appreciated in the community. It is assumed that they are a very high risk group and they are going to die anyway. The second issue is TB. In most of the Eastern and Central African countries, TB has either been steady or going down until a few years ago. Now it's unquestionably rising. One of the things that HIV gives people is tuberculosis. There are a lot of new cases of TB in HIV negative people. There is an atmosphere of despair. A colleague talked with a taxi driver recently who said, "But we'll die before we're 40 anyway." That may be one of the reasons that some of the people were not very interested in getting involved in AIDS research work. The sense of despair leads to a view that nothing is worth starting. It might be interesting to compare AIDS in Uganda to the great sleeping sickness epidemic of the early 1900s. It had a mortality rate of about 30%. The effects on agriculture would be difficult to measure because everyone was moved out of the impacted area in an attempt to stop the infection.

Chin: I would like to summarize some of the global data that I believe would be relevant for the issues of impact on the agricultural systems. The major area that's going to be impacted during the 1990s will be sub-saharan Africa. This conclusion is based on our minimal estimates of HIV infections. The global total of infected adults as of 1990 is at least 8 million.

The growing concern is Asia, but even if HIV begins to increase in Asia in the early 1990s, the number of AIDS cases will not be large until the late 1990s.

We have estimated less than a million cumulative cases worldwide during the 1980s, about five million AIDS cases occurred during the 1990s, and approximately ten million new AIDS cases will occur in the next decade. Depending on what the future level of HIV infections will be, and clearly it's not going to just drop; AIDS cases and deaths will continue to increase. We will see major increases in AIDS cases, and most of that will be in sub-saharan Africa. This is going to be the major area for public health during the 1990s.

Henderson: What's our cumulative total now, again?

Chin: Our estimate is about 700,000 adult AIDS cases and between 8-10 million HIV infections.

Henderson: But publicly you will not quote figures for Zambia - or for other countries?

Chin: No. We will make summaries for sub-Saharan Africa. And we will use three or four central African countries and show the pattern, or use a hypothetical country, unless a particular government gives us permission to release country-specific numbers. I would say with Uganda it is quite possible, because the HIV prevalence data that we're using to model Uganda is based on their official estimate, which is about 750,000 as of mid-1988. They, I think, will not disagree with the numbers, and they want those numbers to get additional support.

Henderson: Jim, this is the first epidemic that I'm aware of in which we have kept track by cumulative cases, rather than annual number of cases every year. Why do we do that?

Chin: For a variety of reasons. One is that if I stopped it, the news media would say, "Why did you stop it?" We put out annual and cumulative numbers. When I first went to WHO they were putting out the reported AIDS cases on a weekly basis. I tried to say that, at a minimum, quarterly. What I wanted to do was twice a year, and I thought even that was too much, because we don't report breast cancer weekly and we don't do it cumulatively. We're dealing with a new disease. The cumulative numbers do have a little more significance than, say, for TB, which is somewhat "stable." It's been around. I think that at a certain point we should give up the cumulative numbers, but whatever I present them I always present annual and cumulative. The most important figure in the future for public health planning purposes will be how many AIDS cases will be alive in any given year since this will be the public health and clinical burden for the year.

Henderson: I think it really distorts the understanding of the epidemic when you show a cumulative total - ten to fifteen million cases a year.

PART FOUR

Nutrition, Environment and Health

7

Food Supply and Health

Michael Osterholm

Osterholm: There have been major changes in our food supply over the last 25 to 30 years. The cardiologists and the public health epidemiologists have convinced us that a meat and potatoes diet isn't the ideal diet to assure longevity. As a result of that there have been major changes in food consumption. Our food supply is much more complex. For example, a well known grocery store owner in the Twin Cities recently commented that when he started his store in St. Paul in 1955 he had approximately 355 items in that store. Today that store stocks over 30,000 different items. That gives you some idea how access to foods have changed. Today I can buy fresh seafood as easily as I could if I lived in Seattle or Boston. The fruits and vegetables that I purchase in our stores here sat in fields of foreign countries seven days ago. Epidemiologist now have to try to understand the international food supply system. May types of fresh food that was supplied by domestic growers have been displaced within the last 10 or 15 year because of cheaper foreign sources. Some have asserted that the industrial sweat shops of the 1920s and 30s have been replaced by the agricultural sweatshops in the 1980s and 90s in many areas of the developing world. It is now cheaper to refrigerate items and bring them to the U.S. from Latin America or from Southeast Asia than it is to produce them in this country.

Let me give some examples related to the issue of surveillance. I am not trying to be outrageous but there is growing evidence that the microbiologic safety of the food supply in the U.S. today is probably at an all time low in modern public health history. We have been grossly over-concerned about things like alar. But we do have tremendous microbiology problems in food supplies that are only rarely being detected. For example just last year a streptococcus outbreak occurred among students at a major U.S. college. The Center for Disease Control was called in. The reason why the CDC was called in was that a week prior to this there had been a sabotage episode in the school kitchen where someone had taken a condom full of feces and put it in the ice cream machine. They thought the two were related. They found, in fact, it had to do with some contaminated mushrooms in cans that had come in from Taiwan. Because of this outbreak, further investigations started to uncover more problems. The FDA has no authority to inspect in any country outside of the United States. They found out that 45% of the canned mushroom coming in actually were illegal. They were not even from licensed manufactures in foreign countries. Had it not been for this episode, in which a saboteur was suspected, this never would have been picked up. We worked on an outbreak last year where members of a professional football team became ill two days after returning from a game in Miami. It turned out that it was due to contaminated food on a Minneapolis based airline.

Ruttan: I thought it was just the bad food they serve on Northwest.

Osterholm: This was especially bad! As we traced the outbreak it involved over 8,000 travellers. It never would have been picked up had it not been for the fact that a large group belonging to the same organization became sick. What it turned out to be was a series of food handling problems in the kitchen that supplied the airline. It illustrates two things. One is the lack of surveillance. The second thing is that 70% of the lettuce that was being used came from foreign countries. Most of it was from Central America or Mexico. Again, the movement of these products are from very diverse areas that are difficult to monitor.

We also worked out the sources of a very large outbreak of salmonella last year. It was a rare type which turned out to be cheese related. There is a major safety problem with the cheese industry in this country. Thirty five percent of the cheese that is being used in this country right now is from foreign sources. It is even coming from places like South America and Southeast Asia, places one usually does not think of in terms of cheese imports. It is being brokered from New Zealand to Thailand and to here. We've picked up incredibly complex sources of food distribution product patterns. Within the United

States, we have cases of cheese that had been produced in Minnesota, shipped to Wisconsin, and sold to a firm in Pennsylvania that, in turn, sold it to a firm in New York, that then moved it back to Illinois, and was ultimately sold in Northeast Iowa. That seems absolutely idiotic in terms of transportation costs. But it illustrates the complicated web we are dealing with.

Finally let me refer to one more case that illustrates the problem. I've had the unfortunate or perhaps fortunate experience of working extensively with a recently documented condition associated with L-tryptophane. It causes a blood and muscle disorder. The connective tissue of the body becomes leather-like. In Spain, 3-4 years after an outbreak in 1980, up to 10-15% of the patients have died. Last fall, our group and a group in New Mexico were able to put indexes in L-tryptophane. We have been able to associate it with a manufacturer in Japan. This particular company produces L-tryptophane using fermentation. We have actually been able to trace it back to certain procedural aspects in the production process. The punitive agent of this particular product appears to be a metabolite of aclymiclophasia. This is a genetically engineered strain that pumps out all kinds of L-tryptophane in fermentation but also pumps out this metabolite.

The reason I bring this up is because of its international dimension. Japan supplies all of the L-tryptophane to the world. The plant is located in the middle of a large petrochemical complex in southern Japan. It looks just like any other refinery. They make this particular material outdoors in a series of tanks. They are very good at manufacturing the organism. We're going to see many other problems that we have never anticipated. This is one of them. And I think it is going to raise issues about genetic engineering in a big way. We did a great job of indexing the L-tryptophane source but we never anticipated that this metabolite was the cause. We actually can tie the problem to strain changes. As they kept getting the strain more and more cranked up to do more and more L-tryptophane the metabolite also went up. And so I think it is going to have long term ramifications.

Finally I would like to emphasize that food system surveillance is something that at best is very poor in this country and almost non-existent elsewhere in the world. We have very little capacity to pick up these emerging problems. And they are very real. I might add for example, for any of you that are eating cantaloupe, there is a big outbreak in this country right now of salmonella chester. We have been able to track it to cantaloupe coming in from Mexico and Central America. CDC has just completed testing on a number of cantaloupe for salmonella chester and as of last week about 30% of all the cantaloupe they were testing right off the market shelves has salmonella.

Straub: Can you give me the proportion of people who are affected by salmonella in the United States? I didn't know it was that important.

Osterholm: That relates, in part, to what Don Henderson was talking about. We don't really know. We are now reporting 85-90 thousand cases a year. This is just the tip of the iceberg. When I look at the outbreaks that we've had here in the state of Minnesota, and try to make those extrapolations, I estimate everyone has one attack every 2 or 3 years. Had the Minnesota Viking football team not gotten sick, we would have never picked up that outbreak. We never would have picked up the cheese case had it not been a rare form of cheese salmonella.

Straub: Has the number of recorded cases gone up sharply?

Osterholm: The number of recorded cases has gone up. But I would be hard pressed to say whether it is an improvement in reporting or a real change. If there is one food setting that we would love to eliminate it is the salad bar. The American salad bar lends itself to all of the worst microbiologic food disease.

Let me illustrate why our knowledge is so poor. We could not identify the salmonella using methods available in American laboratories. Finally we sent it to a researcher in Canada who had worked with a lot of outbreaks in cheese. He found the organism. It was in the cheese packages at the level of about 2 cells per 100 grams. We're talking about an infectious dose of one part in a million. When you are putting 5 million tons of this cheese out in a 4 month period you don't need to have a high level of contamination to get a lot of people sick. That's the kind of thing that does not get picked up. In this country we are too smug about the safety of our food supply. Our problems are compounded in the developing world where I think the food supply itself has to be considered as one source of serious illness along with water sanitation and infectious and parasitic disease.

Ruttan: One of the issues that concerns us in agricultural development is that fresh fruits and vegetables are the most rapidly growing exports from developing countries. This growth has particularly favorable development implications. Production is relatively labor intensive. It is not easy to find things that they can export that are labor intensive. To economists one of the things that we would expect is that we will begin to use our regulatory regimes as a device to protect our own producers. We do have a legitimate regulatory issue. But it has some rather interesting implications for economic development.

Osterholm: In my view as a public health practitioner, and, as someone who has been involved in a number of cases of international transmission, the United States Food and Drug Administration is almost totally incompetent in this area. Under the Reagan administration the agency lost much of its capacity. It is almost an accident when they detect something. The situations I have described were not a result of detecting something in the food supply first but rather responding to an outbreak. I really don't fault the FDA in the sense that they do not have the capacity to do much more. They have had their workload significantly increased and their work force reduced. They don't have the resources to find better methods of detection.

In the case of salmonella in cheese the Canadians were capable of picking it up and the FDA was not. The Canadians have made a real effort in R and D in this area. They have developed methods of detecting salmonella at levels that are now important but that would not have been feasible 20 years ago.

When I came into this business in 1975 we tested things chemically at a level of one point per million. A lot of things looked fine. But now that we can go to one point per billion we find that we have problems. The FDA is still at parts per million in most of what they do - when they do it. They have no jurisdiction outside the United States on food. Even when they were able to go to Japan with this L-tryptophane issue, because it was considered a food additive, they had no legal authority to even look at any of the plants in which it was produced. A lot of countries look to the FDA for leadership and support. But it is a disaster right now. And there are even those within the government willing to say that on the record.

Henderson: The important point that Mike is making is that the surveillance system in this country, let alone other countries, is pathetic. And unless we have a much better surveillance system to measure causes of disease we're going to miss a very great deal. I was particularly struck by James Chin's comment that he is responsible for surveillance and impact assessments for AIDS - and he is only one person. It means we are not yet really concerned about actual measurements, world-wide, of human cases and death.

Allen: Some pesticides that we don't approve for use here come back to the United States in our food supply. As a food scientist I am reluctant to eat tomatoes in the winter. Many of them come from Mexico and I am familiar with their practices. It is a case where countries need to come together with some uniform approach to regulation. I would also like to go back to the earlier question of whether or not there is any indication that the export of foods from lesser developed countries is affecting their own food supply?

Ruttan: It is a very complicated issue because of the potential trade implications. It is one of the areas that has considerable promise from the point of LDC economic development. But we become vulnerable to some of their health problems they become vulnerable to our trade policies. The area devoted to the production of fruits and vegetables for export is rarely large enough to have a measurable impact on domestic food supply.

Osterholm: There have been some serious problems with imported hamburger. As many of you know 99% of the animal that walks into the slaughterhouse gets used -the blood, the eyeballs, and everything else. Methods were recently developed that enabled packers to get an additional 3-4 ounces of additional trim from the gullet. With that trim they were picking up the thyroid. It was going into the hamburger and the people who were eating this hamburger were getting thyroid burgers. This became a source of thyroiditis in this country. It was just a minor processing change. No one would ever anticipate the consequences until we had an outbreak. No inspection services picked it up.

8

Nutrition, Disease and Health

John Murray

Murray: The conventional dogma has it that almost any sort of nutritional intervention will have a favorable effect on resistance to disease, particularly infectious disease. Many years ago David Morley drew attention to undernutrition and its impact on measles. It was accepted by everybody that this was a very important source of susceptibility. Recently Peter Arby in Copenhagen showed very conclusively that there was no real relationship between the state of nutrition and the mortality or morbidity of measles. Instead it was shown to be due to overcrowding. A child goes to school and comes home with measles. In an overcrowded housing situation it spreads to other children. The mortality is more closely related to overcrowding than to the level of infection.

Let me turn to a really basic question. What is the optimal nutritional state, if there is really such a thing, for resistance to infection? I don't think we know the answer. We cannot say that for any group of people or for any individual or for any society that there is an optimum nutritional state in terms of maximum protection against any infectious illness.

I would like to address three issues. One is what happens when people change their diets to resemble the diets in the more affluent western countries. Many years ago, Ansel Keyes, here in Minnesota, reviewed the situation and made the suggestions that perhaps undernutrition, as observed in World War II in concentration camps, was not producing as much disease and infection as people had anticipated.

In about 1974 I had an opportunity - I suppose you would call it opportunity -to work on health and famine in West Africa. One of our striking observations was that during this massive famine, patients in our hospitals developed falciparum malaria. We were providing food to enhance their nutrition. In the surrounding district during this famine, there was no new malaria because there were was no rain and there were no mosquitos. This led us to begin to look at the incidence of malaria and other illnesses in relation to nutrition.

We were able to show at that time that severe undernutrition had an impact on malaria by suppressing it. When we intervened with a feeding program, they developed malaria. Some colleagues got to work on what they thought was going on. John Eaton suggested that the answer was Vitamin E. He noted that the area in which we were working was deficient in Vitamin E as a result of the famine. Vitamin E is an anti-oxidant. Without that anti-oxidant in the diet, the malaria infects the red cells. It imposes a tremendous strain on the red cells and produces free radicals and oxidant stress. As a result, before the malaria parasite can mature in the red cell to the point of infecting another red cell it disrupts. They went to the laboratory and found, as they had predicted, that this particular deficiency of an anti-oxidant had suppressed malaria. What I am getting at is that there are actually some favorable effects of undernutrition in terms of susceptibility to various types of infection. I wouldn't for the moment suggest that people be starved to prevent AIDS. But there is an indication that perhaps there is an optimal intake of food which gives maximum resistance.

I discussed our work with some missionaries who had been living in the same area for years. I asked if they saw malaria patients who came to their hospital for other purposes - whether a broken leg or whatever. Their practice was to treat everybody that came to the hospital for whatever reason with an antimalarial. Our project went on to follow these people for about a 15 year period. We collected data on 4,000 patients with various types of infections. And we convinced ourselves that certain types of infections due to intra-cellular organisms that have to live inside cells are the ones that were suppressed by undernutrition.

At that time we suggested there was some relationship to iron. Iron has become of great interest in relation to infection. Iron deficiency is extremely common in the world. There are various estimates of up to half a billion people involved. I really don't know the exact incidence but it is very common. We had an opportunity to work with Somali nomads several years ago. The Somalian nomads presented a unique opportunity to look at the effects of iron deficiency.

They live on camel milk to a large extent. Camel milk is generally deficient in iron. This whole group of people living along the border of Ethiopia was found to be iron deficient. And yet, they live in an area that has a high incidence of malaria. We began to treat patients with iron for their iron deficiency. Again we observed that the administration of iron to these iron deficient individuals precipitated a number of infections, including malaria and brucellosis, that were caused by intracellular parasites.

We recently presented a paper at a nutritional meeting where we examined immune function tests in people who were undernourished before and after refeeding. There were less infections before refeeding than you would have expected. During refeeding, immune function tests improved but infections increased. Presumably, because of the changes in iron metabolism, there was less oxidant stress on the host cells during refeeding so that microorganisms could survive. We had the paradoxical situation where infections were increasing at a time when they might be expected to be decreasing.

Can we extrapolate this information to societies in general? Is it a useful series of observations? It comes back to the idea that there must be an optimal nutritional situation for each member of a society (or for societies) that give them maximum protection against disease. What happens when this is changed? I think it's pretty obvious, from studies going on around the world, that the long term effects of change in dietary situations - the upgrading to western standards - has had a profound effect on the occurrence of disease. This includes the degenerative diseases which we associate with western society as well as infectious disease.

We have been studying the Maasai for a number of years now. The Maasai traditionally live on milk and blood. They have been extraordinarily free of hypertension, diabetes (type II) and coronary heart disease. We have had the opportunity of following a number of urban Maasai and comparing them with rural Maasai. We studied 47 who had taken a profession in the city of Nairobi. It is hard to find Maasai who are pursuing a professional career in the city. But when you are able to match them with a rural counterpart it is horrifying to see what has happened. The incidence of diabetes in the urban Maasai was over 50% while the incidence in the rural controls was zero. The incidence of hypertension was about 30%. And this can all occur anytime after 4 to 40 years of age in people we have studied. So we are observing the impact of long term changes of nutrition as well as short term.

I don't think anybody could deny that undernutrition of great severity has serious impacts on any group of people, not only from the point of view of work, but from the point of view of survival. Again, I would stress that, in looking at nutrition and relating it to susceptibility to disease, we've got to look not only at the immediate effects but at the long term effects. It is quite obvious that in many societies these diseases are getting out of hand.

Many of you will not probably know the work of Paul Zimmet of Australia on the island of Nauru. Nauru is a small island west of Indonesia. It belonged to the British until 1968. It is an island built on superphosphate. This superphosphate bonanza has made the Nauru islanders incredibly wealthy. The incidence of diabetes, hypertension and obesity had been looked at very carefully before the phosphate exploitation began. Zimmet went back there in 1978 and looked at the dramatic changes associated with the rise in income. They invested the royalties in Australia so everyone was guaranteed an income of $40,000 a year. The food intake for men rose from 3,000 to 7,500 calories per day, and for women to about 5,800 calories per day. Obesity became massive. Diabetes was up to 50%. And worse still, this was aggressive diabetes. It wasn't just the sort we see in the elderly population, which is controllable by various sorts of dietary manipulation. There was gangrene of the hands and other things that you wouldn't see unless you were in a severe diabetic population. Hypertension was present at about 48%. The neighboring islands that had not benefitted from this bonanza of superphosphate were wonderful controls. They experienced only modest evidence of diabetes and hypertension.

Our results suggest that the impact of rapid change in nutritional status may be much greater than among people who have been accustomed to Western diets. Why do these people get so obese? Why were these diseases worse than in our own society? One might suggest that we have had a long time to adapt. I don't think that is the answer. It's interesting to see how the appetite changes in people. For example, in our studies of the Maasai, the caloric intake for a very active male between 16 and 40 was about 1900 calories per day. But when they moved into the city it went up to about 3,800 calories in a very short time.

There is a interesting prototype in an animal model. There is a thing that we call cafeteria obesity in rats. If you put a rat in a cage and let it eat rat chow it won't become obese. But if you take the food from the London University cafeteria and feed it to the rats, they get massively obese in a short time. That doesn't say much about English food. It is a behavioral problem. The same thing happened with our Maasai population. When their diet consisted primarily of milk there didn't seem to be any desire to overeat. If you consumed milk three times a day for 40 years, you probably wouldn't have a great desire to eat excessively. When faced with a huge variety of food there is an incentive to overeat. In a traditional agricultural setting there is little opportunity or desire to eat beyond needs for energy production. It is important that we try and find out for various groups what their optimal nutritional needs are and not encourage them to exceed their needs until their resistance to both infections and degenerative diseases are weakened.

Bradley: We have observed somewhat similar findings among migrants to Nairobi. Blood pressure goes up within 6 weeks of going to Nairobi. One suggestion is the increase in salt intake. My question is whether there is such a thing as optimal diet. There might be a hypothetical population at a hypothetical time. But presumably what will give you the most protection against malaria will decrease your protection against other things.

Whitmore: Do you have any figures on changes in life expectancy and infant mortality - something other than some impressions that they are suffering the ills of the first world and that they might be better off, in some statistical sense, the way they were?

Murray: My own experience has been with the people who have been in a severe famine and were grossly malnourished. Over the last 17 years, we have looked at about 50,000 people in refugee camps. We chose a sample of 4,000 that have experienced a massive amount of malnutrition. We recorded what we observed. It was our impression, in the 4,000 case study that, there was a significant increase in diseases during the re-feeding process. But to extrapolate to a society that is chronically undernourished is very difficult. We don't know the answers to that at all.

Whitmore: You seemed to suggest that they were trading off nutrition for disease. They may suffer from Western diseases. But we do live a long time. And we do have low infant mortality rates event though we suffer all these other aliments.

Murray: There is always a tradeoff. It was our belief that the Somali tribe was living in a nice ecological balance, because of iron deficiency, with malaria, a potentially fatal disease and should that balance change, they would be at a significant disadvantage. This was only a small study. But it suggests a relationship between trace elements and resistance to some parasites. It was a balance that was protective for a whole society.

Allen: Is there any similar data, with respect to any of the helminths that infect the intestine relative to the level of iron nutrition and vitamin B1 or E?

Murray: I recall a zinc deficiency study as done by a group in Aberdeen. Zinc deficiency interferes with the expulsion of certain intestine parasites.

Allen: Animal studies tend to show that undernourished animals in a cage in a laboratory have more resistance in some cases.

Osterholm: I recall an episode in New Zealand where a doctor went up into the mountains and gave newborn infants an iron injection. Then there was a large problem with diarrheal disease.

Murray: Actually, that was a very interesting natural experiment. Iron deficiency amongst the Maori population in New Zealand was very common among infants. The pediatricians had a brilliant idea. Let's give every Maori child that's born a big load of iron. Most of you know that newborn infants have a lot of iron floating around in their livers if they are born full term. Anyway, within a few months, the hospitals noted an outbreak of what they thought was E. coli meningitis. Each of three hospitals in the next few months experienced higher levels of the same meningitis. The pediatricians got together again and said it may be related to the iron. They stopped the whole program and the incidence went back to what it was before.

9

Culture and Nutrition

Doris H. Calloway

Calloway: One of the problems in demonstrating the role of nutrition in disease is that nutritional effects are often very subtle and have very long fuses. That is to say, it may be a lifetime exposure to something in the diet that ultimately winds up increases the possibility of cancer. Also, foods (and diets) are complex, multinutrient and nonnutrient systems, so assigning causality to a specific factor(s) in naturalistic studies is very difficult.

There are powerful barriers to change in household behavior and health-related food practices including resource allocation. One of the problems is the difficulty of not being able to detect the health effects of diet, to develop the individual's understanding of cause and effect. Also, food provides much more than nutrients. It is one of the few sources of pleasure in many people's lives. Humans enjoy sensory stimulation. Almost every culture finds something with little nutritional value but which satisfies a sensory need. Coffee is one example. We like caffeine. We like alcoholic beverages. We like other things that make the day go better. And food has other sensory qualities. It has taste. It has color. It has texture. It has ritual and ceremonial significance. What constitutes a meal is culturally defined. It is not the same from place to place.

Economists usually view consumption of food simply as a response to income. (But when income rises people don't necessarily buy more food--instead of more hamburger they buy steak. Instead of more cabbage, they may buy strawberries.) There is much to learn about what governs the choice between food and other consumption items in different cultural settings. What people do with added money is related to what they perceive to be their most significant deprivation. Our values may not coincide with others'.

Despite the importance of identifying the functional consequences of malnutrition, the data base is deficient. I sometimes think that a problem in drawing research attention to malnutrition is that it is not contagious. Other people's suffering from malnutrition is not widely perceived to affect one's own state. There isn't the same drive at the government level, or even the community level, to try do something about somebody else's malnutrition, as there is to do something about somebody else's contagious disease. Almost no one has enough information to be able to say exactly who is malnourished, where they are, how many of them there are, and exactly what it is they lack or why they lack it. Is malnutrition due, in fact, to bad decision making? To insufficient income? What is the root cause and how can we go about fixing it?

It is difficult and expensive to do the meticulous kind of work that is needed. It doesn't mean just knowing that children are small for their age but finding out interalia what they actually eat and what the food contains. There are roughly fifty nutrients. (A few more trace elements may prove to be more essential than we think now but the total is about fifty.) The goal of food selection is to balance the intake of a few foods to meet the wide array of requirements. Yet, there is little or no modern analytical information on large categories of food items in the world. Dr. Murray believes that malaria suppression is due to lack of iron. It may, in fact, be due to that or to a handful of other nutrients. Controversy will go on forever or until we have the information.

We certainly need better indicators of when people are malnourished and what they lack. This is an area where agricultural research and health research should complement each other. Nutritional considerations need to be built into farming systems research.

We miss an important opportunity by not working together. In traditional American agricultural extension offices there was a farm advisor and a home economist. The home economist talked with the wives about cooking food, how to keep it sanitary, how to preserve products, how to maintain sanitation in the household, and why to put screens in the windows. The early home economics extension agent was, in some ways, a public health worker. There are only a few of those around today. Health care and nutritional services can be delivered together. And home economics extension is not a bad model for the public health nurse in less technologically advanced countries.

Good nutrition practices must become habitual because nutritional adequacy depends on what is eaten every day. You can "fix" an isolated nutritional deficiency by intermittent dosage. Vitamin A is a good example; massive dosages can be given at widely spaced intervals and prevent blindness. Incidentally, the relationship between vitamin A and morbidity from common infectious diseases (measles, etc.) should not have come as a surprise--in had previously been documented in research on Newcastle's disease in chickens.

Murray: It was documented on human subjects in the 1920s and no one paid any attention.

Calloway: That's very true. The point I was about to make is that diets lacking one nutrient are usually short in others as well.
David mentioned vegetables as a source of vitamin A. Why do we have vitamin A deficiency with such a cheap, good source available? There are multiple factors in vitamin A deficiency. Kids don't like green vegetables. Do they know something we don't know about green vegetables? Perhaps there is some reason why green vegetables are not doing the vitamin A job for these children. Fat is required to absorb carotene. Most of the diets are also low fat diets. That may be a factor. At the Asian Vegetable Research and Development Center they are reexamining the bio-availability of carotene in a variety of foods. In the high fiber vegetable crops bio-availability is only about 60% of standard beta-carotene activity. They are now looking at which of the constituents affect the absorption of carotene. The availability is better in commodities such as sweet potato and papaya. Another reason the children have vitamin A deficiency may have to do with the fact that additional bulk is difficult for children that are already on bulky diets to consume.
Diets must have enough fat as well as absorbable forms of carotene and maybe even a little preformed retinol. There are good things to say about eggs in spite of the bad name nutritionists have given them. They are a good food (including being a source of vitamin A) and many children would be better off if they had more of them.

Ruttan: Please comment on the major set of nutrition studies you directed in Egypt, Kenya and Mexico.

Calloway: I was involved in the planning and management of the three projects that made up the USAID-Sponsored Collaborative

Research program on Nutrition and Human Function (NCRSP),. but the projects were directed by binational teams of investigators from the U.S. and host countries. These studies are unique in that food intake was measured 2 days a week for a year in about 300 households in each location. Consumption was recorded for the household and individually for the lead male and female, a school child and a toddler. Each of the projects has reported its general findings but more definitive analyses are still underway.

Our group at Berkeley also received funding to carry out analyses access-projects; these are in progress. For our analyses which attempt to link outcome variables (growth, pregnancy outcome, disease experience) to intake of energy and nutrients we have had to create a food composition data base. This International Minilist (IML) has a relatively small number of foods (less than 400) but contains complete entries for each of about 50 nutrients and related factors. The foods included are the most widely consumed staple crops and items that serve as proxies for a group of foods of similar composition (e.g. apple is listed in the IML and is a proxy for pears, white cherries, etc...). No reliable analytical data exist for some foods that were consumed. These were classified by appearance (e.g. a leaf is light, medium or dark green), or source (e.g. marine or freshwater), or other observable characteristic. Estimation of nutrient intake from this base, while less than perfect, does allow us to make a reasoned judgment about the probable adequacy in the diets as consumed across an entire year.

Clinical symptoms indicate an advanced state of deficiency. There were few such symptoms recorded in the NCRSP populations; these were more commonly noted in school-age children and pre-schoolers than in adults. Adults can withstand things that children can not--such as parasite burdens--that affect nutritional status, but anemia was fairly common in all age/sex groups except male adults. Preliminary analyses of nutrient intakes (based on the IM) suggests that several nutrients are low in each group and these nutrients are not alike in all locations. Some--vitamin B_{12} in Kenya, for example--have not been recognized as community health problems. The next steps are to explore the possibility that nutrient sources have been missed and to look for evidence of deficiency (such as blood levels) in the population.

There are other interesting topics yet to be explored from the NCRSP. Our Mexican study was drawn from area that produces pulque. Pulque is a mildly fermented beverage which is about 2% alcohol. (It's made from cactus juice, a food source that is otherwise not utilizable. It's not like taking perfectly good corn and changing it into whiskey. Cactus juice has an array of carbohydrates that are not absorbable by humans. When the juice is fermented it changes something that is not exploitable into a source of nutrition.) Traditionally, it is used daily as a source of calories. In the Mexican study, it accounted for about 10 to 12% of calorie intake. These people are not fat. They need the calories. So that is a benefit. The study area is in economic transition. Nowadays the men often go to Mexico City to work 5 days a week and come home to the village on the weekend. Instead of consuming some pulque each day, some consume a large amount in two days, leading to drunkenness. It seems likely that nobody eats as well on the weekends. Inebriated parents may not look after children so well. This is an example of unintended effects of economic change that may not easily be detected.

Rosenzweig: In the two discussions on nutrition I have heard echoes of a debate that development economists had in the past concerning agriculture. And that was the issue of whether, within the context of traditional agriculture, experts had anything to offer in the way of policies that would improve productivity. I have a similar question in the area of nutrition. In areas where traditional knowledge provides little guidance, but in areas where there has been little change for many years or many generations, do we really think we have the superior information that your extension worker can provide? Should we be surprised when we see that a family feeds working members better than infants? You were saying on the one hand we don't really have much information but on the other hand you were suggesting that there would be some gains from sending in an extension worker to tell families what to do about nutrition.

Calloway: One assumes the extension worker comes from the same area and has access to the traditional knowledge as well as scientific knowledge. Generally speaking, advice in nutrition has been given on the basic of generic recommendations that are generally sound - such as that people should eat from a variety of food classes.

Rosenzweig: I understand that model. But assume that you go into an area of India in the semi-arid tropics where people have consumed the same food for many generations. Do you want to go and tell them to change their diets?

Calloway: Not without knowing that they have a problem. You need to know whether or not they have a problem.

Rosenzweig: Suppose they are malnourished. They are very poor. At their level of resources can they be more efficient in their consumption? Do you think that we have the knowledge to tell them to change their diet given their same level of resources? There are no new resources and no new technology. But the idea persists that we can somehow go in and tell them how to change their practices to become more efficient. I think most agricultural economists would agree that in the absence of technical change or changes in markets there is very little to tell them about production. Why isn't that true in the area of nutrition?

Bell: Mark, I have sung this song myself for many years. But it's not only a question of applying pre-existing knowledge. It is also a question of applying knowledge that is discovered on the spot by technical people who have been trained to observe and analyze. Doris, is there evidence that people, when given modern scientific knowledge, have changed their consumption? Ken said a little while ago in some places people don't eat papayas. It would seem very odd if over the centuries they had not discovered that papaya is a good thing to eat. Is there a lot of that or is it by and large the hypothesis that Mark and/or I would use - namely that they have adapted pretty darn well to what is available to them given their income?

Calloway: The person deciding isn't necessarily the one who has the problem. With vitamin A deficiency it's mostly the children who have the problems and it is the adults who decide what is eaten.

I want to return to Mark's question. You said the people are poor and they are malnourished. But deciding whether malnutrition has been present is not easy. The fact that children are under-height or -weight for their age does not necessarily indicate malnutrition. It is an outcome that may have had multiple causes.

In the Egyptian population that was studied, there was generally no food availability problem, nor were children generally underweight. Yet mortality among infants was higher there than in the two less well-nourished populations (Kenya and Mexico). One of the nutritional factors in Egypt may have been that the mothers give sugar water to children from almost the day they are born. Even though people heat water for their baths they do not necessarily boil it before they drink it. The infants started off at reasonable weights and by the time they were 6 months old they had already fallen behind the standard markers for both stature and weight. They were the sickest children among the three studies. There was much more diarrhea and much more respiratory disease.

Rosenzweig: It almost sounds as if you are saying you don't know anything.

Calloway: To know how much of an outcome is due to "malnutrition" you must first know that food intake is limiting. Many studies have been done in populations, particularly in the Americas, in which most adults are not malnourished. That being so, it is unlikely that food supply per se is a limitation. It is more likely to be a question of food quality or bulk-density. I'm saying we don't know because we haven't looked very carefully. We know children are often stunted. We know families are generally poor. But that is not enough. We need to know what they lack.

Warren: When I got started in infectious and tropical diseases there was a major theory that malnutrition was a basic underlying factor of disease in the developing world. It was the malnourished children who then became more suspectable to respiratory infections and to diarrhea. It was generally held that nutrition was the basis of the decline in infection in Britain over the last 100 years. Everybody ignored the sort of thing you are talking about. Then I became an immunologist. If you think of increase of susceptibility to infection you have to think in terms of immunology. What I learned from the immunologists was that only severe malnutrition has an effect on cell mediated immunity. I was very confused at this point, until Leonardo Mata in his Children of Santa Maria Cauque (MIT Press) revealed in village studies in Guatemala that infection was a major cause of malnutrition. Maybe you nutritionists know more about the situation than I do but I think that Mata really changed global perceptions about the role of nutrition. He convinced me that, at least in this modern world, infections are a more important cause of malnutrition than malnutrition is a cause of infection.

10

Water and Disease

Conrad P. Straub

Straub: I will abbreviate my discussion since David Bradley has already touched on some issues that I might have covered. I share his perspective that it is time to think more seriously about rural water supply problems than we have in the past. The WHO programs started out concentrating on the urban areas first. The urban areas have been under continuous pressure to catch up with water supply needs because of rapid growth. In the larger urban areas water supply and treatment facilities are very similar to those in developed countries. This is often a problem because, due to a lack of trained maintenance people and operators, it was difficult to provide the chemicals needed for treatment and to keep the sophisticated plants in operation.

In the rural areas it is possible to make large improvements with simple methods. There is a center at Delft, The Netherlands, where information is gathered on simple water supply and treatment techniques found useful in developing countries. By building on indigenous methodology, some very unique schemes have been developed that are applicable to particular areas.

Some of the best experience in developing rural water supplies has come from several South American countries where local people used indigenous materials and resources in developing their systems. Since these were locally developed projects they were maintained and operated more efficiently and effectively. In such systems, as in all systems, water should be supplied on a continuous rather than intermittent basis. In India technicians have been specifically trained to monitor and maintain wells. Water supply sources are visited by bicycle-equipped personnel on a routine basis. The spare parts are carried and repairs made as needed.

The problem of disposal of sanitary wastes is more difficult. Provision of facilities does not automatically result in use or maintenance. Cultural differences must be considered, as well as the fact that human excreta may be used for fertilizer. More complex wastewater treatment systems often stand as "monuments of disuse," because of a lack of adequately trained technical and operating personnel. Although they require fairly large land areas, the use of oxidation ponds for wastewater treatment, when properly maintained and operated, have proven effective. If properly designed and operated, the final units of the system can be used as fish ponds thus increasing food resources.

Another important area is food distribution. This includes transportation resources, proper refrigeration during transport and proper handling and storage at distribution points. It is unlikely that many developing countries will be able to maintain the "cold chain" that links food production, processing and distribution in developed countries. Irradiation, with gamma rays, might offer a partial solution if emotional considerations could be overcome.

As a result of the Chernobyl reactor accident with its widespread release of radioactive contaminants, local as well as far distant agricultural areas became contaminated. This had a marked effect on the local consumption of foodstuffs that were contaminated as well as wide repercussions resulting from the export of radio-actively contaminated agricultural foods. The same would be true when there would be a release of toxic chemicals that would contaminate the land and be taken up by the agricultural products grown thereon.

Bradley: The points that Conrad made raises several interesting general issues. He mentioned villages building their own water supplies. I think one big problem is the traditional view was that government would build and the villages would maintain. In the last decade people have reversed their perspective. The villages are good at building things because it requires a short period of intense activity which can be timed to fit appropriately. But you need a bureaucracy to maintain the system. The second point was about the dramatic increase in health benefits is when you pipe the water into household, not just to a neighborhood stand pipe. If you do that by indigenous methods it can be done at a reasonable cost. You need the tap in the compound to get water usage to the level where you get substantial health benefits. There is an enormous amount of effort being put into trying to document the health benefits of water supplies.

The results are not terribly conclusive for the amount of effort. It is quite interesting that the research has now shifted to looking at the time benefits to the women rather than using the health problems as a primary rationale. The last issue is that of the use of feces for a variety of purposes. It seems much easier to handle the health issues if you are dealing in a culture where feces are used as a resource, as in parts of Southeast Asia and Vietnam, than where they are treated solely as a waste product.

Ruttan: Why? Can you elaborate on that point?

Bradley: In some societies feces are a waste product that no one wants to know about. To suggest that something should be done to make them safe is very difficult. If you are dealing with a culture which has traditionally used feces, not as a waste product but as a valuable resource, it is easier to convince people they should store them for two weeks before using them.

Henderson: I was wondering about the issue of quality and quantity of water. The issue has certainly been around a long time. We have been trying to demonstrate that one does get a difference in terms of illness if you supply pure water but it has been difficult to demonstrate. The suggestion has been made that quantity may be more important that quality. If this is correct we are wasting our resources attempting to improve the quality of water. Where does our knowledge stand at this point?

Straub: It is a very reasonable argument. The first approach is to find water. Then use the best available sources that you can find. Then you try to make the needed investment to get access to the water. Only later should you become concerned about quality.

Bradley: The whole literature about water quality was created essentially on the assumption of the problem of quantity had been solved. The western literature primarily addresses the issue of quality. But it didn't address the water availability question because that was solved in the last century or earlier. Since there were disease sources that were water bourn it made sense to limit assuming water quality. But in developing countries the main route of transmission is by water. You should do a lot of other good things before you get into the question of water quality. I don't want to denigrate water quality. But it is a matter of priorities.

Allen: David, is there is a need, on this quality issue, to make a distinction between the needs of infants and children compared to the population as a whole? Doris has mentioned this issue. When we talk about high mortality of diarrheal diseases isn't that significantly related to water?

Bradley: It's related to water. But in developing countries much of it is transmitted by the lack of water rather than through water. If there is no water for washing then you can get very high levels of person-to-person transmission. One can not over emphasize availability.

Warren: David Bell mentioned earlier, when we were discussing malaria, that it is the infants and children and not the adults living in that environment that are killed by it. The same may be true for a lot of water borne and bacterial infections. You see local people brushing their teeth in water from a dirty canal with apparent impunity. But if you brushed your teeth in that water you would be sick as hell.

Bell: The issue of the effect of nitrogen fertilizer on water supply keeps puzzling me. I keep hearing that in the United States there are very serious problems in certain parts of the country. Is that relevant to what we're talking about here?

Allen: In the infants stomach the pH is neutral. The microbes in the stomach take the fertilizer nitrate and reduce it to nitrite or nitrous oxide. That firmly attaches to the hemoglobin and then the infant is incapable of transporting oxygen in its blood. This is known as the "blue baby" syndrome". This is a very serious problem whether the nitrate comes from fertilizer in the water or in food consumption. That's the only issue in terms of fertilizer.

Ruttan: Dave, you referred me to Gordon Conway and his work. I was surprised at how little health impact they found in both their water and pesticide studies in Asia.

Straub: I would like to make two final points in regard to water quality. We refer primarily to meeting the bacteriological standards based upon the presence or absence of the coliform group of microorganisms. There can be some leeway in meeting the standards recommended by WHO and promulgated by the EPA. There may also be differences in the so-called secondary or aesthetic standards. Higher odor, color, turbidity, or salinity concentrations can be tolerated easily as can currently accepted values for sulfate, iron, total solids, etc. So that is what we are referring to when we say that quantity may be more important than quality.

Where bacterial limits established by local authorities are exceeded, concentrations found in the raw water may be reduced by boiling, disinfection, or the use of slow-sand filters. The latter require very simple maintenance and function well when operated properly.

Warren: I'd like to make some comments on water and agriculture. Schistosomiasis is intimately related to the building of dams and irrigation systems. It's also related to swamp rice cultivation in West Africa where enormous levels of transmission of schistosomiasis occurs. I think anytime that anybody is involved in agricultural projects that involve water management, particularly irrigation, it is essential that they consider schistosomiasis and ways of mitigating its transmission.

Another problem is the water-related, insect-transmitted diseases, because whenever you impound water you get the breeding of insects, in particular mosquitos, but others as well. Black flies that transmit onchocerciasis are actually encouraged by dams, because the black fly grows best in rushing water.

Osterholm: This gets back to is the issue of surveillance. I'm very skeptical of the significance of nitrate or nitroglohemia. There is no question that it can occur. But if you go back to the work in the early 20s and 30s and into the 40s there was apparently a very significant problem of anemia in this state. In Southeastern Minnesota the area is underlain with limestone. There is good communication between the surface and the wells. We continue to have a problem down there.

It's improving because now people have deeper wells. But there are many, many families having water consumption histories of high nitrate and high bacteria levels. Bit I'm not aware of a single case of nitroglohemia in this state in the past 15 years with one exception. It is puzzling why we don't have more problems.

11

Health and Environment
in Central and Eastern Europe

Zbigniew Bochniarz

Bochniarz: First of all, I would like to make a significant distinction between geographical and political terms -- Eastern Europe. If we look at the map of Europe, it is easy to notice that a little more than half of European territory belongs to the Soviet Union. This is of cause the Eastern part of Europe. From that point of view such countries as Poland, Czechoslovakia, Hungary or East Germany cannot be also located in Eastern Europe. From a geographical point of view these countries belong to Central Europe (Middle Europe). However in everyday life in the West these countries are still listed as East European countries. This is due to political conditions created after WWII according to the Yalta agreement, which condemned these countries to the USSR dominance zone. Fortunately for these countries, the recent revolutions have made them independent from the Soviet Union and in such circumstances there is no reason to call them East European countries. They also do not want to be called that way anymore.

This historical process means also that these countries are no more a military threat for the West. They present however another threat for the West and the rest of the world -- the ecological threat. If we look at the map of Europe showing concentration of sulphur dioxide, the areas with highest concentration -- 1000 or more micrograms of SO_2

per square meter -- are all located in Eastern (Donietsk -- Ukraine) and Central Europe (Southern Poland, North-Western Czechoslovakia, and Central and Southern parts of East Germany). These areas also belong to the part of Europe with the highest acidification of precipitation and soil. The threatening thing is the growing territory of these regions and their integration into one large region of environmental disaster in the heart of Europe. It is already the most polluted region not only in Europe but in the world. (Besides the above mentioned countries, there are other countries with very high acidification of precipitation and soil such as Hungary, West Germany, Luxembourg, and Scandinavian countries.)

If we take into account other kinds of emissions to contributing either to global warming or to acid rain -- CO, CO_2, NO_x -- the East (USSR) and Central Europe contributes a lot to European and global emissions. From that point of view any progress achieved in reduction of greenhouse gases or acid rain emissions in the West (Western Europe and the USA) can be easily offset by a constant increase of these types of pollutants in Central and Eastern Europe. From that point of view we need to collaborate in order to achieve a progress in the global scale. This is also an interesting topic for a collaborative research program.

The environmental situation in Central and Eastern Europe looks even more interesting when we compare emissions or deposition of major pollutants per capita or per unit of GDP. In the first case the major Central European countries reached the level of the most industrialized countries. If we compare pollution per $1,000 of GDP in the middle of the 1980s, the Central and East European countries even overpass the major Western countries. In emissions of particulate matters per unit of GDP, the six Central European countries (Bulgaria, Czechoslovakia, East Germany, Hungary, Poland, and Romania) emitted 13 times more than the EEC countries, in the case of SO_2 their emissions per unit of GDP was 2.5 times higher than in EEC-12. These facts are quite significant and ironic (pervasive) in the context of a permanent political and ideological mobilization in these countries over the last 40 years to reach and overpass the most developed countries.

Explaining factors behind these indicators, one should take into account the structure of national economies-- both sectoral and spatial -- natural resources and conditions, strategies of development in the past and functioning of national economies. There is no doubt the natural endowment in these countries is poorer than in the West (less caloric fuels, poorer soil, harsher climate, etc). It explains partly their higher energy use per unit of GDP as a substitution of these less convenient conditions. But these reasons cannot explain sufficiently, why Hungary and Poland have 3 times higher energy intensity of GDP than Japan or 2 times higher than that of the EEC countries. The major reasons for such high energy intensity in Central and Eastern Europe are associated with their development strategies based on heavy industry and central planning as a major tool of implementing these strategies. In such circumstances with obligatory plan targets and artificial price structure, there was a lack of interest in efficient use of energy and natural resources which in turn, caused high level of pollution.

What is the impact of such pollution on health in that region? As you probably well know, for many years it was impossible to obtain any environmental data form this region. Recent political changes released some of the interesting data and results of still limited research related to environmental and health issues. They show that adverse impacts on human health are growing in the most polluted areas of CEE. There are several demonstrable features of the most polluted areas in terms of human health and life expectancy. In the late 1980s, mortality rates (especially those caused by circulatory diseases) in the region of CEE were the highest in Europe (Figures 11.1 and 11.2). There is growing evidence of higher morbidity rates due to respiratory diseases, cancer, and circulatory diseases among the inhabitants of the most polluted areas in the CEE. Figures 11.3 and 11.4 provide graphic evidence of the differences in average life expectancy at birth and at age 45 in CEE in comparison with the West.

There is also a very interesting phenomenon of growing difference between life-span of men and women in Central and Eastern Europe. Recently it is over 7-8 years. The average life expectancy at birth for a male is about 65-66 and it is declining or stagnated. There is also a stagnation in life expectancy of female at about 72-73 years in the region. Let me now look more closely at the relationship of environmental quality and health indicators.

Poland started publishing environmental data in 1972. In the middle of the 1980s, this data was supplemented by maps of environmental quality -- the first such maps in the region. The first presentation of these maps was a kind of shock for the Polish society. According to this data, 27 areas were listed as areas of environmental hazard by the Polish Parliament in 1983. Five of them -- Gdanska Bay, Glogow-Legnica-Copper Mining Region, Cracow Region, Katowice

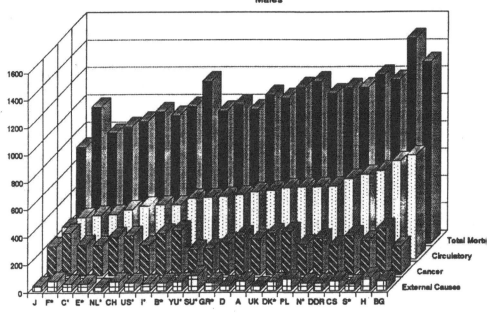

Figure 11.1
Standardized Mortality Rate per 100,000
Males

Data is from 1989; * = 1986; ' = 1988.
Source: '1990 World Health Statistics Annual,' Geneva 1991.

89

Figure 11.2
Standardized Mortality Rate per 100,000
Females

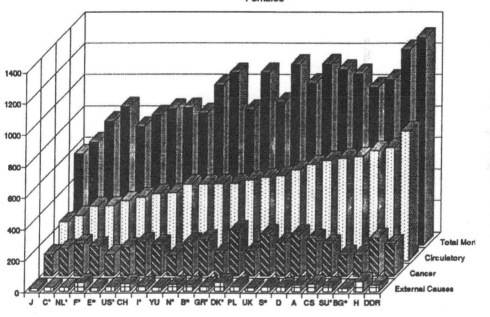

Data is from 1989; ° = 1986; ' = 1988.

Source: "1990 World Health Statistics Annual," Geneva 1991.

90

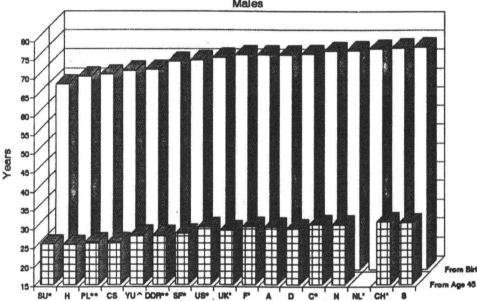

Figure 11.3

Comparative Average Life Expectancy
Males

Data is from 1988; ^ = 1985; * = 1986; ' = 1987; ** = 1989

Source: "Demographic Yearbook," United Nations 1989.

Figure 11.4

Comparative Average Life Expectancy
Females

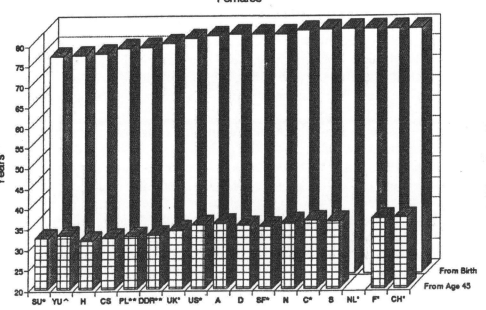

Data is from 1988; ^ = 1985; * = 1986; ° = 1987; ** = 1989
Source: "Demographic Yearbook," United Nations 1989.

Region and Rybnik Coal Mining Region -- are officially recognized as areas of ecological disaster. These 27 areas cover about 11% of Poland's territory inhabited now by 35% of the population.

The official definition of environmental disaster indicates that all ecological balances are broken in the area. The declining ecosystems have lost their capability for natural recovery. For these reasons human intervention is needed to rebuild the shrinking ecosystems. From a practical point of view these areas are characterized by closed beaches, poisoned wells, resettled villages due to soil and water contamination, high content of heavy metals (the world record was noticed near Katowice where cadmium contamination in soil reached the level 20 times higher than that described by the WHO as risky for human health). Besides these areas there are in Poland 50 heavy polluted cities (usually with both air and water polluted), which jointly with the areas of environmental hazard are inhabited by more than 60% of the population.

These data show also an interesting and dangerous trend. At the beginning of the 1980s, about 10% of Polish rivers possessed drinkable water (first class of water quality according to the Polish norms). Five years later, in the middle of the 1980s it was only 6% in the first class. Recently issued data of the Main Statistical Office in Poland does not show any major river in the first class according to biological criteria. As you see, these changes over the last 8 years are very dramatic with far reaching consequences for the human health and environment. You should also know that in Poland, like in East Germany and Czechoslovakia, the surface waters are major sources of drinkable water. Unfortunately the water quality in these countries is no better than in Poland.

Recently Czech colleagues from the Charles University in Prague developed environmental maps dividing the country into four class. The last -- fourth class -- could be compare with the Polish category of environmental disaster area. There are significant parts of the Czech Republic listed in that category in Northern Bohemia and Northern Moravia regions. Slovakia -- the less developed republic -- does not have many of these areas, except a part of Eastern Slovakia around Kosice. Comparing these environmental quality data with health indicators such as infant, male and female mortality, and life expectancy in developed and underdeveloped areas it is hard to offset the impact of much better health service in the first category of regions. But it is interesting enough that the second highest infant mortality rate -- after highest mortality rates among minorities (Gypsies and

Hungarians) in Slovakia -- was noticed in the most polluted ares in the Czech Republic. It was also quite obvious correlation between pollution and female life expectancy. In other words, the factor of higher quality health service was visibly offset by pollution and environmental degradation.

According to research published by the Warsaw University over 40% of monitoring stations in Czechoslovakia, East Germany and Poland registered higher concentration of SO_2 than the upper limit of the WHO (45 micrograms of SO_2 per m^3 during 24 hours). In the case of NO_x the violation took place in about 62% of stations and in the case of particulate matters respectively 82%.

What are the health consequences of such development. In addition to what I have already mentioned, it should be emphasized that recent life expectancy for males between 40-60 of age fell to the level of the 1950s. In other words, after progress in the late 1950 and 1960s and at the beginning of the 1970s, there has been a significant drop in this very productive part of the society in this region. There are, of cause several factors contributing to this dramatic phenomenon, but I am deeply convinced that one of the major factors has to be pollution.

Environmental hazards are affecting not only the urban population but the rural one as well. About 33% of the Polish soil is heavily acidified. In Czechoslovakia it is about 45%. And in East Germany it might be even higher than 50%. In addition to acidification the overdosed pesticides use in Hungary, East Germany, Czechoslovakia, and in some parts of intensive agriculture in Bulgaria and Poland, is a common phenomenon endangering living conditions in rural areas, which are still inhabited by about 44% of the population in these countries with the working population in the agricultural sector still very high -- between 20-27%.

There are some differences in case of death causes or morbidity between rural and urban population. The rural population suffers more from stomach ulcers and digestive cancer -- mostly due to water pollution. In the case of urban population the typical causes of death or morbidity are respiratory and circulatory deceases.

As I have already mentioned, the pollution and environmental degradation is one of the major factors contributing to high mortality and morbidity rates in Central and Eastern Europe. Besides that, we should remember that these countries are among the to heavy drinking and smoking countries. There is also a lot of stress caused by political and economic crisis, shortages of basic goods, job security, overwhelming environmental threats and others. Recently, we discussed in Poland an interesting phenomenon of the decline of the immune barrier due to stress. This way all kinds of pollution is getting an easier access to the human organism and causing more sever damage to it. Other words all these factors create a kind of synergy effect accelerating mortality and morbidity rates.

Osterholm: The decline in life expectancy in recent years is very interesting. Did GDP per capita also decline over that period?

Bochniarz: There are many controversies related to GDP measurement. There are always differences between the World Bank, CIA, and UN estimates of the GDP of Central and East European countries coming out based on different methodologies and objectives. According to my research, the GDP in this region was almost always overestimated. In the case of Poland the mistake was very significant -- between about $2,000 per capita according to Encyclopedia Britannica, to $1,700 per capita according to the IMF in 1988/9. I think that it is about $2,000-2,200 now. In the middle of the 1980s the estimates were quite close -- between $2,500 -3,500 per capita according to different sources.

A common phenomenon for the whole region has been stagnation or even decline of GDP per capita. In the case of Poland it dropped about 30% between 1978 and 1982, and then stagnated.

Straub: The data on this disaster you have described is going to be of the greatest value for all of us as we look more carefully at the effects. It will be very valuable to get careful data on congenital malformation, pulmonary malformations, lead contamination and others. What we're suddenly discovering about lead is frightening indeed. I suspect around some of those plants there is a lot of lead in the water and in the soil. Measurements of effects on learning ability, growth and so forth may be very valuable.

Bochniarz: Thank you for this question and suggestion. I think that we would like to share our data with the rest of the world. We are also ready to undertake some joint research in order to help better understand the metabolism of some processes as well as to help other countries to avoid our mistakes. We would like to disseminate the results of such research like the Japanese did in the case of Minamata disease. So far, we have discovered that due to heavy metal pollution the share of retarded children is significantly higher than Poland's average in the Silesia Region. They also differ from the rest of the country in hemoglobin composition as well as significantly lower weight of newborn babies. Miscarriages are meaningfully higher there than in the rest of Poland. In terms of preventive action, the government is granting the Silesian children one month of vacation outside of the region annually.

Herdt: I'm convinced of the absolute level of the problem. But the data that you referred to on the decline in quality of water in the rivers during the 1980s was incredibly rapid. What kind of industrial pollution or other factors could have changed so rapidly in a ten year period?

Bochniarz: There are several factors contributing to the above mentioned dramatic changes in the water quality in Poland. The data of 1988 describing water quality are based on biological and physical and chemical criterion. According to the first criterion, there are no longer any major rivers with drinkable water during the so-called "sugar campaign" (this is the period of major discharges from seasonal sugar plants). According to the second criterion, after the campaign period only 2% of rivers could be classified in the first class. Obviously, even under the second criterion the situation does not look much better than under the first. I think that we have reached a kind of threshold beyond which the deterioration is progressing very rapidly. I have observed this process of accelerating deterioration of water quality in Polish lakes in the Northern part of the country during my last five years of sailing. In other words, when the accumulation of pollution reaches a certain level one can expect a significant accelera- tion of deterioration in ecosystems. And this is what, I am afraid, is taking place in Poland now.

Osterholm: What are the criteria you use in classifying water as drinkable and undrinkable water? Could you tell me what criteria you use to classify the rivers?

Bochniarz: 1987 was really a relatively dry year. This explains, however, only a small portion of water pollution. But how can one explain the growing emissions of major pollutants faster than the rate of GDP growth in the 1980s? Water consumption also grew faster than the GDP. And so did pollution. During the GDP decline of about 27% in 1979-1982, we observed a significant increase in SO_2 pollution at an annual rate of over 3%.

There are two criteria for classifying the quality of rivers -- a physical and chemical criterion and a biological criterion. The last criterion is based on measurement of coli bacteria and a saprophage index. The physical and chemical criterion includes, e.g., biological oxygen demand (BOD), chemical oxygen demand (COD), suspended matters, and heavy metals. There are three classes of water quality according to Polish standards based on both criteria: drinkable water (Class I), water suitable for agricultural purposes (Class II), water suitable for industrial purposes (Class III), and water beyond any class (wastewater).

I do not know of any city in Poland where you can drink tap water. You must always boil it. The situation is nothing like it is here. Only in rural areas do farmers drink directly from their wells, despite the fact that in about 60% of wells the water quality is poor -- below the standards. Besides organic components in water, there is a high level of salinization in tap water in the Polish major rivers -- Vistula and Odra. It comes from coal mines in Southern Poland. Recently salinity was measured at 15 times higher than some 30 years ago. Our water plants are helpless under this phenomenon. They are doing quite well with organic pollution but not with salt.

PART FIVE

Disease, Health and Development

12

Historical Population Collapse

Thomas Whitmore

Whitmore: Conventional wisdom has it that the global total human population has been growing exponentially - upward and onward from the cave. In a study that some colleagues and I did a couple of years ago we examined regional-scale populations over millennia and found that regional-scale populations fluctuate a great deal in contrast to the global-scale total population. That is not surprising. In the cases we examined, there were near-millennia-long periods in which population grew substantially, often doubling or more, but there were also long periods of considerable population decline, to one-half or less, in all of the case studies. These "waves" of population growth and decline all predate the modern "population explosion." We are not sure in every case what caused the declines or what triggered the expansions, but health-related problems are probably responsible in several of these cases. Population growth and its concomitant pressure on resources may have contributed to the population retreats in several cases as well. While these ancient growth rates (on the order of 0.25 to 0.5 percent per annum) are not comparable to modern rates, they may have had the same sort of disruptive effects on local economies and ecologies in ancient times that the 3 to 4 percent annum modern rates currently have in parts of the Third World. Agricultural productivity declines, perhaps caused by environmental degradation, have also been implicated. (Johnson and Whitmore, 1986; Turner, 1986; and Whitmore and Turner, 1986).

The second point I would like to make is based on my current work. As you are all probably aware, there was a large population of indigenous peoples in the New World in 1492 and a great many fewer one-hundred to two-hundred years later. Nevertheless, there is a good deal of controversy about exactly how many Amerindians there were in 1492 and even about how many there were one-hundred years later. For these reasons the scale of population collapse is also uncertain. I have been concerned especially with the Amerindian population collapse of sixteenth-century Mexico.[1]

I approach this issue differently from the way it has usually been examined. Traditionally, historical demographers have used Spanish tax-roll counts and other documents dating from the latter half of the sixteenth century to make estimates of the total population at those dates. Earlier-date populations were estimated from these later-date estimates by back-projection - literally drawing a line on a graph using an assumed rate of population collapse. (Borah and Cook 1960; Cook and Borah 1960; and Cook and Simpson 1948). I am much more interested in what caused the population collapse. To that end, I have used a systems dynamics computer simulation methodology to model the sixteenth-century population "system" of the Basin of Mexico.

The goal of this simulation is to model the complex effects of the series of epidemics that affected the Basin in the sixteenth century. This model incorporates explicit demographic assumptions and a model of the Basin's agricultural production systems; both interconnected with a model of the epidemiological conditions of the sixteenth-century Basin. My simulations indicate that a 90 percent collapse is a reasonable estimate of mostly epidemic-induced mortality for the first 100 years of European occupance in the Basin of Mexico. Famine accounts for 10-15 percent of the population loss in this period. Relatively speaking, it is minor compared to the catastrophe caused by the epidemics. The thirteenth-century Black Death in Europe killed "only" 25-45 percent of the population in a 100-year period (Hatcher, 1977:68; Slack, 1985:15; McEvedy and Jones, 1978:42; Russell, 1948: 260-281). But, in the New World, where the entire population was at risk to each of the European diseases, there were three Black Death equivalent epidemics, one after another - smallpox, measles, and bubonic plague. Typhus and influenza also took their toll.

---

[1] A good introduction to this vast literature may be found in: Crosby [1967]; Crosby [1976]; Denevan [1976]; Gibson [1964]; and Hassig [1985].

The key point for our discussion here is how the introduction of these exotic diseases in the Basin of Mexico affected the population's ability to feed itself. It is widely recognized that famines and epidemics are co-travelers. There has been some controversy about which precedes the other. Nevertheless, it is clear that they interact. By building a simulation model that allows me to more carefully control that relationship, I can model the effect that a given epidemic will have on labor availability, and therefore, the effect it has on food production. By reducing labor availability, epidemic mortality and morbidity can lead to food shortfalls and subsequent famine resulting in more morbidity and mortality.

My model of the disease-agricultural production interaction was necessarily simplistic. This is a complicated issue and it is hard to model it realistically. Further, it is better to keep simulation models simple to avoid the confounding effects caused by too many interacting variables (Randers, 1980). Nevertheless, doing this modeling forced me to come to grips with a variety of issues on the interaction of health and agricultural production.

It is important to consider the nature of the organization of agriculture. A key is to recognize the local agricultural ecology, technology, and labor needs of particular farms (Turner and Brush, 1987). This is really a repeat of Doris Calloway's plea for on-farm research and of David Bradley's concern for local effects. The first consideration here is the mix of production types in the region of interest. Small agricultural enterprises typically fall on a continuum ranging from production mostly for household consumption to mostly for the market. Most Third World farmers fall in the middle - they produce for the market and they produce for their own consumption. This point is important because farmers often practice very different types of management on those two parts of their farms. Health issues related to the use of modern inputs such as fertilizers, pesticides, and herbicides may be more important for farmers who practice market agriculture because they are more likely to use these inputs on that part of their production.

The availability of transport, bulk processing, and markets all affect farm profitability - and farm profitability has a direct effect on the health of the farm family. The ability to afford things such as better housing or clothing may make a subtle, but noticeable, impact on health in the long run. Further, the availability of infrastructure such as all weather roads probably is associated with the easier provision of public health services as well. On-farm storage, food processing techniques, and cooking facilities form a third agriculture-health related issue. They directly affect food safety and thus, rural health. Agricultural development that addresses these issues helps improve rural health at the same time as it increases agricultural production.

Straub mentioned seasonality effects, and these are important as well. When are people sick? Is there seasonal malnutrition? It is important to recognize episodic or seasonal morbidity and how it relates to the agricultural labor cycle in any planned agricultural development. Farmer health impacts agricultural practices according to the timing of labor inputs that are necessary in the particular agricultural system. In places where there are seasonal diseases, reduction or elimination of them may pave the way to much increased agricultural production without any other changes. But, agricultural development schemes that do not take into account or remedy seasonal illness patterns may fail. Further, agricultural development schemes that alter the local disease ecology such as forest clearing or irrigation or alter the timing of necessary labor inputs may alter the pattern of illness and thus adversely or benevolently affect output.

Related to these seasonality effects is the issue of who is sick. Who does the agricultural labor around the farmstead, and what kinds of labor they do, is critical to understanding the relationship between health and agriculture. It was mentioned by several people that the health of children affects agricultural production, since mothers are frequently responsible for both agricultural chores and child rearing. A sickly child at home means less time in the fields - and presumably a smaller harvest. Similarly, work in the fields may inhibit the proper nurturing of sick children and raise the possibility of increased mortality. Further, high childhood and infant mortality rates are associated with high birth rates. In many societies, women do a great deal of the agricultural labor. Thus, more frequent pregnancies may disrupt agricultural labor as well. Women who are pregnant are also more vulnerable to a variety of health problems including malnutrition. The last point about who is sick concerns the possible trade-offs between rural and urban health. Agricultural development that raises production may benefit the urban population's health by making more and cheaper food available while at the same time that it diminishes the health of the rural population. Who benefits and who pays as a result of development is not easy to determine and possible trade-offs may involve people far beyond the local development scheme.

A different time scale issue was raised by David Bradley - the long time-scale effects of any sort of agricultural development on health.

Any sort of change that alters the demographic structure of the agricultural population may change the ability of those in the productive ages to support their dependents. This point was borne out by my simulations of sixteenth-century Aztec times. It was quite clear that the early incidence of epidemics lead to drastically altered dependency ratios later in the century. By the middle of the sixteenth century, the productive members were supporting far more dependents than they had been traditionally used to doing. The literature of the period refers to labor shortages, and the difficulty of finding enough able-bodied men or women to do the necessary agricultural work (Hassig 1985: 178-185). Further, circumstances that push or pull family members, particularly males, off the farms may reduce the amount of agricultural labor available. Thus, even urban developments that encourage migration may have a detrimental effect on rural health.

Warren: With most diseases, the people that are killed are children. Elderly people also may be more susceptible. This means that you would probably have more able-bodied people, as a percentage of the total population, than before such an epidemic.

Whitmore: I would argue that that's not the case in a "virgin soil" circumstance. Everyone is susceptible. No adult had ever had smallpox before when smallpox arrived in the Basin of Mexico. So adults are equally susceptible as children. Ordinarily, you're absolutely right.

Ruttan: I want to come back to the dependency ratio issue. Africa is going through an AIDS plague right now. It's hitting a very specific part of the population. It is killing those who are at their most productive age. This will exaggerate the effect on the dependency ratio.

Osterholm: The key factor with AIDS that distinguishes it from all the others is that it is sexually transmitted. The others were mainly respiratory and that makes a big difference. The only one comparable would be syphilis. But of course syphilis' impact was minimal relative to what AIDS will be. We have to understand that AIDS effect will be very different than we have ever had before in terms of plagues.

Whitmore: Except maybe in the 16th century in Latin America.

The other change that has a similar effect for a range of diseases is labor migration. For example, migrant laborers from northeast Thailand who go to southwest Thailand die of malaria there.

The other thing that makes AIDS extremely difficult, besides wiping out this population that is going to be the most productive for the next 20 years, is that they leave behind large numbers of orphan children. There's a need, within the system, to find a support mechanism for that unique group. We've never had a disease before that had selectively left large numbers of children with no parents to support them. If the mother is infected, the father is also likely infected. They will both die. That's unique among major infectious diseases. The agricultural question is, what difference does it make. Have we made enough gains in agricultural practices so you can take out 20-40% of the labor and not affect food production?

Calloway: Inevitably, children will move more rapidly into the labor force. In which case there will be less opportunity for education and less opportunity for technological advancement. I don't know now where altruism is going to come from for housing and educating the children.

Bell: It strikes me that a lot of our recent conversation has been surprisingly anecdotal. We are coming on these questions as if they are novel. We are a relatively sophisticated group, which suggests that these are not issues that have received fairly steady and serious attention. I'm not talking about the 16th-century Aztecs, but who's affected by worms and what happens culturally when there are a lot of handicapped kids around. There have been serious episodes--is somebody following the Bhopal consequences, for example? Are we stumbling over a kind of blind spot in public health, or in scholarship more broadly? It seems to me these are extremely important issues, and yet we don't have very much solid historical analysis, follow-through of analysis, and so on.

Chin: You have pointed again to another blind spot. Our concern with AIDS is leading to the discovery of all sorts of deficiencies in our health infrastructure and our research. During the reign of Idi Amin, within a six month period more young adults were killed than AIDS will probably kill in a 10-year period in Uganda. Where are the demographic studies to look at the implications in Africa?

Whitmore: The population was increasing during the worst of the Amin years. The population was increasing at virtually a maximal level.

Chin And it still is. But where are the studies? Questions are being asked about the same thing.

Henderson: We have so many kinds of adaptive mechanisms that they are difficult to sort out. But are we missing something from past events? The great Leningrad famine during World War II might provide evidence on how people adapted. And there is a lot of data on that.

Rosenzweig: There is a very well-known phenomenon in Africa that is known as child fostering. It's an old tradition. In African countries, there are these large and extended structures, in which children will cross households when there are too many dependents or too few dependents. One of the things the World Bank has done in the last couple years is promoted and paid for a lot of surveys in African countries. One of the things that turned up is the importance of fostering. What's being studied right now is the extent to which household and person specific emergencies or problems are being dealt with by this shifting. Households respond very quickly to emergencies, from person-specific illnesses to droughts. There are some anthropological studies in South India. The Caldwells lived in a village where, for two years in a row there were no crops. But there was very little out migration and not very high mortality. The answer was contributions from outside the village from relatives. So these extended relationships across households in dealing with emergencies are very important. They're just beginning to be studied.

Ruttan: One other thing that strikes me, in some of these AIDS areas, is that agricultural production is particularly labor intensive. Furthermore, the possibilities of substituting animal power for human labor in parts of tropical Africa are quite limited because of animal disease problems. This means that the flexibility of this system may not be nearly as great as in a system where there is either animal or mechanical power.

Warren: I'd like to bring up worm infections in children again because this goes along with what David Bell was talking about - areas that could be of immense importance if we had reasonable data. The data we have over the years is on the very high prevalence of infection with helminth parasites. Many individuals are infected with several different species simultaneously. But the quantification of infection has been difficult. David Bradley was one of the people who played a major role in reviving the whole idea of quantifying these infections. Many papers published still only have prevalence data. Without doing microfilaria counts the data are essentially useless. We now realize that another confounding factor is the over-dispersed distribution of the helminth infections. Only a very small proportion of people, say about 10%, have heavy infections. Those are the ones who get sick.

Unfortunately, school age children tend to have the highest levels of infection. Many of the outcomes do not occur while they're school children, but when they reach adulthood they develop elephantiasis, river blindness, and hepatosplenic disease. The infection rates go down in both prevalence and intensity in the older people. In the present context this information is of importance because many school age children are old enough to work on the farms, and a significant proportion of them will be chronically ill with the above mentioned diseases.

13

Agricultural Development

John Sanders

Sanders: In considering agricultural development and health issues in Sub-Saharan Africa, it is first necessary to eliminate the gloom and doom of much of the popular literature which emphasizes accelerating population growth and stagnant or declining per-capita agricultural production.

In spite of the poor aggregate agricultural output growth in Sub-Saharan Africa, there have been regions of rapid development as, for example, in some higher rainfall regions of the Sahel (Savadogo, 1990). With new technologies to increase water availability and soil fertility, there is substantial potential in other lower rainfall regions (Sanders, Nagy and Ramaswamy, 1990). Moreover, with the sequence of good rainfall years in the second half of the 1980s, there is reason to be optimistic about whether the recent droughts imply a climatic and vegetative shift (decertification) or only reflect a cyclical climatic phenomenon (Gray, 1990).

There appear to be linkages between investment in health and agricultural development besides the obvious inference that healthier workers should be more productive workers. Let's consider two cases.

Semi-arid West Africa can be divided into three agro-climatic crop-production regions based upon rainfall and resulting vegetation (Table 13.1). Moving from south to north, approaching the Sahara, the rainfall levels go down and cropping systems change. In the high rainfall cotton region (Sudano - Guinean), the diffusion of new

TABLE 13-1 Characteristics of the Major Agro-Climatic Zones in Semi-Arid West Africa: Rainfall, Cropping Systems, Population Densities, Technology Introduction, and Technology Development.

Agro-Climatic Definition	Rainfall Levels[a] (mm.)	Total Area[b] (%)	Population[c] (%)	Rural Population Density (People/Km^2)	Cropping Systems	Crop Technology Adoption	Population Density	Crop Technology Development Potential
Sudano-Guinean	800 - 1100	24	6	9	Cotton, peanuts, Sorghum, millet, vegetables, cowpeas	Rapid. New cotton cultivars. Fertilization of cotton, maize, Sorghum. Animal traction and microtractors.	Moderate. (High disease risk)	High
Sudanian	600 - 800	21	59	20	Sorghum, millet, cowpeas, vegetables	Some. Countour dikes for water retention. Animal traction - donkeys. Early cultivars of sorghum and cowpeas.	High	Moderate
Sahelo-Sudanian	350 - 600	30	19	19	Same as Sudanian, but moving farther north; millet, cowpeas, and nomadic grazing.	Dikes and donkeys. Farther north, minimal.	Low (low rainfall and more fragile soils.	Lower than Sudanian for sorghum system.

a. 90% Probability

b. The other agro-climatic zone in semi-arid West Africa is the Sahel with less than 350 mm of rainfall (90% probability).

c. The other agro-climatic zone in semi-arid West Africa is the Sahel with less than 350 mm of rainfall (90% probability).

Sources: Rainfall divisions are from J.E. Gourse and D.R. Steeds, Desertification in the Sahelian and Sudanian Zones of West Africa, Washington, DC: World Bank, 1985, p. 2. For more detail on the cropping systems, technology adoption, and potential technology introduction, see J.H. Sanders, "Agricultural Research and Cereal Technology Introduction in Burkina Faso and Niger," Agricultural Systems 30 (1989): 139-154. Also see J.L. Posner and E. Gilbert, "Sustainable Agriculture and FSR in Semi-Arid West Africa: Keeping the Elephant Out of the Rowboat," mimeo presented at Farming Systems Association meeting, Michigan State University, East Lansing, MI, Oct. 1990.

technologies, including new cotton cultivars and chemical fertilization of cotton, maize, and sorghum, has been rapid. The French made substantial investments in agricultural research, infrastructure, and marketing agencies in this region.

This high-rainfall region with many fertile river valleys has not only a more favorable agricultural environment but also is a more favorable environment for the malaria mosquito, tsetse fly, and the black fly, which carry river blindness, than the two other crop regions (Table 1).

Historically, the Sudanian region has been a more healthful environment. Population density is higher in the Sudanian zone than in the Sudano-Guinean zone. Technology development is more difficult. With the elimination of the long fallow period as population density increased, soil-crusting has become more serious, further reducing water availability to plants in a region in which the absolute levels of rainfall are already low and the variation in rainfall between years is substantial. The simultaneous introduction of technologies, both to increase water availability and soil fertility is more complicated for adaptive research and for farmers' managerial requirements than the technological requirements in the Sudano-Guinean zone involving improving and maintaining soil fertility and introducing new cultivars.

Migration has been increasing in the last two decades from north to south, gradually extending the population pressure and resource degradation into the higher-rainfall regions. To slow the rural-to-rural and rural-to-urban migration and to increase farmer incomes, there have been large expenditures of foreign aid over almost twenty years to develop new technologies for all three zones, but especially for high population-density regions in the Sudanian and Sahel-Sudanian zones, such as the Mossi Plateau. In contrast, the Sudano-Guinean zone has much lower population density, more resources, a more dynamic environment resulting from previous technological change, and more potential for technological change (Table 1). More investment to reduce the environmental health hazards in the Sudano-Guinean region should facilitate more intensive settlement and further agricultural development. Clearly, an integration of health and agricultural technology planning is required to further develop this higher-rainfall region.

A second question is what is happening to agricultural incomes and technology introduction in the river valleys that have been "cleared" of river blindness and then resettled? The river valleys generally have more agricultural potential than the more settled uplands which historically were outside the range of heavy black fly infestation. The disease cycle apparently has been broken with the chemical control of the black fly by the Onchocerciasis Control Program involving spraying of the breeding sites supplemented with the use of the Merck drug, Ivermectin, which breaks the cycle of the parasite in the infected human.

In some "cleared" river valleys in the Sudanian zone after more than a decade of settlement yields are falling. The role of cotton and fertilizer is decreasing, and real incomes have decreased by 40% over the decade based on comparisons of two normal rainfall years. Farmers have returned to extensive soil-depleting practices. Many of the better farmers have left the area. In this Sudanian region, without investment in new agricultural technology, the long-term payoffs to the successful health investment in river blindness control have been declining.

In contrast in the "cleared" river valleys of the Sudano-Guinean zone farm incomes are high and apparently increasing over time as fertilizer use increases, especially on cotton and maize. The clearing of these valleys in the Sudano-Guinean zone is turning them into boom areas. New agricultural technology was already available for this higher-rainfall region. A marketing infrastructure, first for cotton, and then for maize has been put in place. Clearly, health and technology development investment activities are complementary and need to be evaluated as linked investments.

Bell: Are there other factors that could be involved in this, such as external aid, that stimulated the boom after getting rid of the onchocerciasis? Are there any entrepreneurs who brought up large plots of land and are then hired people to work for them?

Sanders The initial process of colonization in the older area was traditional colonization where everything was pre-planned and people were even paid to move. However, the Solenso migration in the Sudano-Guinean zone was spontaneous and largely unplanned. Some of the companies in the Sudano-Guinean zone are now providing incentives for farmers to use microtractors and guaranteeing the purchase of corn. But it is mainly spontaneous migration. And most of the migrants are the Mossi, who come out of the poorest area of Burkina. I don't think that there are factors other than that this is a prime agricultural area for which productive new technologies have been developed.

Rosenzweig: I'm confused by your story. Maybe I'm getting bogged down in the geography. Are there three locations you're looking at?

Sanders: Just two sites.

Rosenzweig: Both of which are cleared of disease?

Sanders: Yes.

Rosenzweig: And there was migration to both of these sites?

Sanders: That's right.

Rosenzweig: One boomed and the other didn't? Both areas had settlements, both areas were cleared of the disease.

Sanders: The older area in Sudanian region was cleared first. Both areas were cleared of the disease at different times. You have an older area now ten years old where there were substantial benefits from in-migration, as evidenced by the income comparisons between the donor and recipient communities and by the rapid spontaneous migration around the planned migration areas. But after ten years of following an extensive agricultural development strategy with fertilization, only on cotton, yields and incomes have declined. People started to move out. There were substantial benefits from the initial development in this older region. And it appears that there was a substantial multiplier effect from economic growth in the region. But after 10 years, without new technologies to stabilize and increase yields and incomes have declined 40% in real terms.

Rosenzweig: So you're making the case that the benefits would have been larger if a more appropriate technology had also been introduced?

Bradley: What was the ex-ante situation in the two areas?

Sanders: In the older area there was very little settlement except away from the river valleys, in the marginal lands before the river blindness control program. With the clearing of river blindness, people moved into the valleys which are the lands with the most potential. In the other area, in Solenzo, there was some settlement before the area was cleared of river blindness because these are really good soils and people were willing to take some risk.

Rosenzweig: You're not showing anything about onchocerciasis or non-onchocerciasis. You're just showing that there were two different approaches to the resettlement of these areas -- one which seems to be causing more rapid development. But they were at different time periods so they're not controlled for that.

Sanders: When you compare the migrants' income in the doner region, to their incomes in the new regions there were substantial income differences. Without the control for oncho, people would not have moved into these areas and the production would not have increased. In addition I am trying to show you something else. I am arguing that there is an interrelationship between the gains from disease eradication and the development of the new technologies. If you don't invest in developing the new technology then you're not going to get sustained development - even if the development effect resulting from disease control is initially successful.

Henderson: His main point is that health or agricultural programs considered in isolation do not necessarily lead to long term benefits. In one instance it was done poorly, and in the other instance it was done well. It seems to me it's a very rich and interesting example, supporting the argument for combining health and agricultural planning, which is the main point.

Rosenzweig: Do you have any information on what happened to the areas the migrants came from?

Sanders: Yes, we have some data from other sources but not from this survey.

Osterholm: Did incomes rise or go down?

Sanders: They're not rising. They're stagnant or declining in the doner areas.

Osterholm: But what was the effect of the out-migration on those areas? Do the residents who do not migrate experience benefits from the fact that other people have gone?

Sanders: Only from the remittances, in both money and in cereals, from the river valleys to the donor areas.

Rosenzweig: But there are no gains in productivity as a result of lower population pressure in the donor areas?

Sanders: We haven't been able to measure it. This was a study conducted by a team of sociologists and anthropologists and we were not able to get everything that we might have if it had been designed for our purpose.

Ruttan: Mark, you are asking for conclusions from a study that wasn't carried out.

Warren: I'm concerned because I have had a great concern for the quality of information. It is my belief, from my bibliographic work, that probably no more than 1% of the studies are worth anything. I'm absolutely serious. You're laughing--I don't know if it's nervous laughter. You think I'm crazy. The problem really lies both with the original experiment design and the quality with which the study is carried out.

Sanders: The case points up that when you use studies for different purposes than for which they were designed you may not be able to answer all the questions you would like to.

14

Health, Nutrition and Work

Mark Rosenzweig

Rosenzweig: There's been a lot of discussion throughout the day-and-a-half that we've been here about the lack of data. We've heard a lot of questions raised. Some are old hat. Some are new. I want to talk about findings. They're findings from studies, some by me, from general purpose surveys that have been collected in the last decade or so in low-income countries. The growth of these general purpose surveys from low-income countries has, in my view, revolutionized development economics.

In recent years, these general purpose studies have been expanded to include topics that generally were not previously thought of as being important. One of these areas has been health. So these general purpose surveys are unique in combining detailed information on demographics, on economic relationships and, at least in recent years, on some aspects of health, nutrition and food consumption. They're extraordinarily large. Many are national probability samples. They provide, for the first time, information about the incidence of various illnesses and of nutrition across a population.

The questions that I want to address with data are familiar but I want to pin them down. First of all, how important is morbidity in these populations? Is ill-health a problem just in local areas but not a national development problem? Secondly, does health improve with economic growth? The aggregate data that we have suggests that it does. Countries that have developed have gotten healthy and countries that haven't tend to be less healthy. There are similar kinds of correlations with fertility. But do we have any better evidence? How is economic development gong to affect health? Third, do we know anything about interventions, given the level of development that can affect health? Fourth, how does health affect productivity? What are the costs of ill health to societies?

We immediately see that the fundamental problem, once we have the data, is dealing with two-way causation. At the same time that we're asking if economic growth improves health we're also asking if health affects economic development. It's going to be very difficult to sort this out. The studies that I'm going to talk about are studies that have been sophisticated enough to recognize the problem and to do something about it. I obviously won't have the time to talk about what they did and none of what they did is perfect because there are no controlled experiments here. There are just general purpose surveys. In economics what we do is substitute econometrics for experiments - which is the same as saying that you substitute assumptions for experiments. Sometimes the assumptions are reasonable and sometimes they're not.

Let me first give you some information about morbidity. I've reviewed a lot of surveys by health economists and others about health in low-income countries. When you look at the data you find that they have described the general problem - you find mortality information. It's interesting and important but it isn't the same as health. Of course, it is very difficult to measure health. I decided that it might be useful to take some of the general surveys that I had available to look at the issue of what proportion of the population say they are sick. I'll give you a little more information about what sick means in these populations. I have four data sets: Indonesia, Ivory Coast, India, and the United States. So the United States is a benchmark. Because of the configuration of these data I was able to get a little more information out of some than others. Indonesia did a national probability sample in 1978 of about 3,000 households. The data show that 17 percent of the households all over Indonesia, had some person sick the previous last week. The average number of days sick for these people was 16 days. They had been sick two weeks when they were found to be sick in the previous week. Eight of those days were in bed. In fact, 75 percent of those who were sick reported they were sick enough to have to be laid up in bed. About 2.3 percent of heads of households were sick in the previous week. For the U.S. data 0.8 percent were sick. So it's four times as high in Indonesia, if the question means the same thing, which we don't really know.

In the Ivory Coast we have a probability sample of about 13,000 individuals - 1,600 households. They asked about sickness in the last month. Of individuals, 31 percent reported they were sick in the last month. The average number of days they were sick, for those who were sick, was 11 and 6.7 of those days were in bed. About 71 percent of those who reported sickness were sick in bed. Half of those, by the way, received some kind of care, mostly public care, and 32 percent received hospital care. I looked at the age distribution. Among males, ages 25-40, 35 percent reported some illness in the past month. The mean number of days lost for the ill was 9.4 in the month with 5.2 in bed. If we blow that up to the total population, that's an average over all prime-age males in the Ivory Coast of two days of sickness in bed per month per adult. That seems like a lot. that seems like something to worry about. It seems like this is an issue.

In India we have five villages. There was a probability sample within the villages. the villages are supposed to be representative of the semi-arid tropics of India. They are a pretty large population - 717 adults. Twelve percent reported they were sufficiently sick in the last year that it prevented them from working. The proportion sufficiently sick so they couldn't work for the landless households was 17.2 percent. For the largest land-owning group, those with 20 acres or so on average, it was 7.8 percent. So more than twice as many of the landless were sick than the upper landed group. It suggests something about the distribution of illness in the population. It suggests to the extent that sickness prevents work and work drives income that illness exacerbates inequality in income in the population. This is not a new idea but now we have some information that it exists. So there seems to be an issue. There seems to be a lot of illness, however illness is defined.

The second question is does health improve with economic growth? From our literature survey we find studies by economists that look basically at the effects of income on food consumption and on nutrients. Generally what is found is that the effects of income on food consumption and nutrients is very small. The income effects on food consumption are greater than the effects on nutrients - not a surprise. As people's income goes up, they tend to eat more expensive foods but not necessarily get more nutrients out of them. The effects are non-linear. The effects of income on nutrients and food consumption are much higher at the lower income levels. It is discouraging that you get such small effects, because to relate those to the time-series information we have is very hard. One of the big problems in economics is that the micro-cross sectional results are not easily related to the time-series movements.

There is a notion, mentioned by Calloway, that culture very importantly constrained what people do. But what's generally found in the literature is that when food prices change, as they will with development, there are significant shifts in food composition and some, but smaller, shifts in the nutrient composition as well. There are also important distributional effects. It's obvious, for example, if you look at rice farmers, that when rice prices increase, that those people who tend to produce for the market benefit. And those who tend to consume mostly their own rice lose. So you get some strong effects. The little evidence there is on illness information and prices finds a mixed bag - no general conclusions - except one which is that a rise in the price of sugar increases nutrition. So sugar is bad - but I wouldn't put too much weight on that.

This issue about the relationship between cross-sectional results and time-series is illustrated with findings from another study. The green revolution in India provides, in some sense, a natural experiment because it occurred only in selected areas of the country. It therefore becomes possible to compare changes in various things in those areas with the changes that occurred in other areas of India. What one finds is that where the green revolution occurred there was a decline in fertility and an increase in school enrollment compared to areas in which it didn't occur. These effects are much stronger than what you pull out of the cross-section. But it does suggest that there is rapid response on the part of household allocations, inclusive presumably of health, to technological change in agriculture. On interventions, given levels of development, there is even less information. There is some evidence from these studies, consistent with what we heard before, that piped water does reduce illness. Proximity to health clinics reduced infant mortality, but it reduced it significantly less for well-educated mothers. Now this is a finding that's intriguingly similar to what's found for the effect of extension on agricultural development. It's often found that the effects of extension on agricultural development is much smaller for the well-educated farmers. And it suggests that these clinics may be providing information that the well-educated mothers already have. In fact, there's a very strong regularity in the data, which is that maternal education and child mortality are very strongly negatively correlated. That's robust to measures of income. And no one fully understands why that's true, and I would hesitate to start pushing up levels of education for women on the basis of that relationship. But it's robust across very many countries.

Bell: That's a first approximation policy conclusion.

Rosenzweig: That's true. It certainly follows logically. And I wouldn't be against it for other reasons. Now what about productivity effects. These are most interesting. There's only been two studies in the entire world history that attempt to take into account the problem that income may affect health and health may affect income. That was done in a Sierra Leone study. I think it's a good study. What was looked at was basically estimated agricultural production functions, so you have all the normal inputs for agricultural output. But what was added was the average calorie availability to the household workers in the survey. And what was found is that a 10 percent increase, at the mean in availability of calories, increased output by 3 percent, controlling for time and other inputs. For families for which the average calorie consumption for adults was about 1500 calories, a 10 percent increase in calories would lead to a 5 percent rise in output. For the 5000 calorie availability, that doesn't necessarily mean they're consuming that amount, a 10 percent increase only increased output by 1.2 percent. It's non-linear as we would expect. So it again suggests that the poorest groups benefit most from economic growth, and you get the greatest impact on output in these groups as well. There is no evidence that well-fed people in these environments received higher wage rates in the market. So all this comes from household productivity. That remains a bit of a puzzle to economists. In fact, we have some very grand theories based on the wages that are tied to nutrition for which there is no support.

There are consequences not only for home productivity. As has been mentioned a number of times, there are reallocations of resources in an economy when someone becomes ill. The question is, what are those reallocations of resources? We have some information on this. Again, these are studies from national probability surveys. These are not isolated villages or ethnographic studies. These results are coming out of these large-scale surveys which generalize, in principle, to the population. The Sierra Leone study was not a national probability survey but it was a probability survey.

Henderson: These surveys were all conducted in the same manner?

Rosenzweig: No. These are different surveys from different institutions.

Rosenzweig: There is an issue of third party effects. An interesting study looked at the effect of illness of a child in the household on the allocation of time for everyone in the household. Again, this study took into account the fact that the allocation of time affects whether a child is ill. The interesting, and perhaps surprising, result is that in Indonesia the major effect of a child being ill was not on the work time of the mother, but was rather on school attendance of the teenage girls. That is to say the teenage girls were substituting for the mothers in caring for the child. Their time was being pulled from school. So what illness of little kids tends to do is exacerbate the differences in human capital investment between girls and boys in Indonesia. That seemed to be one of the principal effects, by as much as 20 percent, when the child was ill.

There's a large literature on the effects of fertility on labor supply. It turns out that the major effect of having small children in the household depends on whether or not those children are ill rather than their mere presence. There's another kind of third party effects that people are just beginning to study. That is the issue I raised earlier about transfers across households. There appears to be, in many populations, a lot of money moving across households--across long distances. And these transfers appear to be responsive to income emergencies in households. The question is whether they're also responsive to illness. Now if they are, it means that the effects of illness, or illness eradication, cannot be evaluated even if there's no migration, just in the area alone, because it means that resources from other areas, remittances and transfers also are going to be affected. So you have to be very careful about looking at the total impact of any kind of illness reduction.

The final generic issue is whether or not the allocation of foods in households, that is, the intra-household allocation of foods to individuals, which is controlled by households--at least some members of the households--responds to incentives. There was an interesting study in India, this one based on a national probability sample, which could explain a large part of the differential infant mortality of boys and girls by variation and opportunities for adult women to work. This has been replicated, a number of times. In cotton growing areas which employ a lot of women, the differential is much less in favor of boys than girls, compared to the mean. In Muslim environments, you get much bigger mortality rates for girls compared to boys relative to the average than you'd expect. Of course we know boys, on average, have a higher mortality rate than girls and it's not just cultural.

In Bangladesh, more calories are allocated to family members who are engaged in energy-intensive activities. This is not a surprising result, but it explains in part the significant average differential between the calorie consumption of men and women in Bangladesh, which everyone knows about. Adult men on average consume more than women. But it also explains the second moment in the distribution, namely that the variation across men in calorie consumption is much greater than it is in women. And the explanation again seems to be that the activities of men are much more varied than they are for women in Bangladesh, in terms of the energy expenditure of those activities. And because one seems to relate to the other, you get this big variance. So these are some of the findings from these kinds of studies. Now the general surveys obviously have the problem that everything is not done well. But the fact that you can relate a lot of things done reasonably well to each other gives you, I think, some insights into what's going on.

Warren: I'd like to start out with the whole question of the definition of health and illness. The standard definition of health that I have on the wall outside my office in New York is a poster from the World Health Organization that defines health as complete physical, mental, and social well-being. And they mean it! John Knowles held a meeting and published a book called <u>Doing Better and Feeling Worse</u>. And that is a phenomenon, not only in the United States, but in many of the other developed countries, where there's a lot of money. There's no question that if you measure parameters of health, we're doing much better in terms of both mortality and morbidity. There was a study written by Barsky about three years ago in the <u>New England Journal of Medicine</u> in which he presented specific data, going back to the early part of the century, that there has been a growing disaffection with the state of people's own health and with the health care system.

What about complete physical, mental and social well-being? Black did a study in England in the 1970s in which he showed that the greatest single cause of encounters with the medical establishment at every level, from primary to tertiary health care, was mental illness. As physical illness decreased mental illness became a bigger problem. A recent book has compared the enormous cultural differences in medical attitudes in the U.S., U.K., Germany and France. The French have, as we all know, an obsession with the liver. The Germans have an obsession with the heart. Nevertheless, life expectancy is similar in all of these countries.

The last thing I have to say is that, if you're talking about health and economic development, there may be little relationship between GDP per capita and life expectancy. We held a meeting, <u>Good Health at Low Cost</u>, in Bellagio several years ago. China, Sri Lanka, and Kerala (India) with GDPs per capita of $300 had life expectancies of about 65. Costa Rica with a GDP of $900 had a life expectancy of 72. Thus, the poorest can have good mortality statistics overall. Improving health is possible without depending on much money from the outside. It is worthy of note that Saudi Arabia with a GDP of $7,000 had a life expectancy well below any of the above countries.

Rosenzweig: Let me talk about the definition of illness and then about this issue about good health at low cost. There's no question illness is a subjective measure. And there's no doubt that it will differ across cultures. I have no question about that. It will even differ within a country. There is this strong negative relationship between maternal education and infant mortality. We thought we'd find the same thing with illness. And we found just the opposite. It was clear that the more educated mothers were reporting illness more frequently because they cared more about it. There's no question that shows up, and of course we controlled for it. Within a country, you can make some inferences about what determines illness--across countries you probably can't. But the best method--and people are beginning to implement that because the data are becoming available--is to look at the same person over time.

One of the new areas of survey is of course longitudinal data which involves following families and individuals. I think those kind of data are going to be really valuable.

Now, on the issue of interventions versus economic growth - and it's a very important issue - the micro surveys don't have much to say. The income effects are small. The intervention effects are small. That's probably a measurement problem. But the idea that good health can be had when incomes are low is right. It's the low cost part that I think is wrong. When you look at the expenditures on the part of the governments in those areas they were very, very large. Costa Rica doesn't have an army. They were allocating a large amount of resources to the health and education and they were getting a pay off. It is good, but it is not low cost! We had two people from Costa Rica at the Bellagio meeting who were jumping up and down saying "it's not low cost." You will recall we had a great fight there, because some people wanted to call the meeting "Relatively Good Health at Low Cost." We decided that would not be a compelling title for a book. So I think it suggests that you can, with a lot of resources, have an impact on health even with low incomes. The question is whether that reallocation towards health was one of the causes of low income in those societies wasn't addressed at the meeting. In the case of Karala, they are paying a price for the way they're extracting the money to pay for health and education. What the appropriate balance is, I'm not sure.

Warren: But there are very few resources in Sri Lanka and Karala. Even if they put a very high proportion in health it would still be a very small amount of money.

Rosenzweig: But suppose they'd put that money into agricultural research? This is the crux of the issue about health and development. Do you put the extra dollar or rupee or whatever into interventions in health? Or do you put the extra dollar into technology development and adoption? Where do you get the highest payoff? We really don't know. We know in the Karala case and the Sri Lanka case that the economies are performing far below their potential. That may have something to do with extracting the amount of resources for health. We don't know.

Bell: If I understood you right, Mark, you said on productivity that there were two studies. I only heard you describe one.

Rosenzweig: There are two studies that deal adequately with this problem of reverse causation. The second study was in India. It used essentially the same technique except for the distinction between calorie consumption and measures of nutritional status. There was a significant relationship between weight per height of the family worker and productivity. But the wage relations are not believable. In Sierra Leone there weren't any. So I think that has not been definitively pinned down.

Bradley: In saying there are only two studies, what did you mean - two studies using this particular type of methodology? There are loads of studies.

Rosenzweig: No. Two studies whose methodology is scientifically sound.

Bradley: Do you mean scientifically sound or useful? There are quite a substantial number of studies.

Rosenzweig: But most are nonsense. They don't take into account food consumption for example. There was a study done in which they gave food supplements to cane workers and watched how much more cane they could produce. But they didn't take into account what they were eating at home. They didn't even take into account the reallocation among household members. There have also been other studies that looked only at the simple associations between what people eat and productivity.

Bradley: Are you saying there are just two studies that examine the relationship between eating and productivity, not between health and productivity. When you say there are only two studies what relationship are you considering?

Rosenzweig: The problem is that income will affect what people eat and their health will affect their income. The studies must deal with that reverse causation problem - the identification issue - to be believable. When I refer to only two studies, I mean studies which didn't suffer from the problem that we can't tell whether it's income that's increasing their consumption or their health increasing their income.

Bradley: There are quite a lot of studies showing that you can improve health by specific interventions. You can treat the cane cutters for schistosomiasis. The best way to deal with reverse causation is to intervene and see what happens.

Rosenzweig: Absolutely. But you've got to be careful about the linkages in the experimental design. As I said, a lot of this econometric analysis is trying to substitute use of survey data for the controlled experiment that you'd like to perform but which of course you can't perform on large populations. Certainly you're not going to reallocate food in the household in an experiment.

Henderson: The studies you cite are a rather thin basis on which to make broad generalizations. What they do suggest is that there might be some virtue in getting some health people together with some economists and doing some studies. I think there is scope for looking at these questions, both investments and outcomes, with more elegant studies. We do have the World Fertility Survey which does collect a reasonably consistent set of data over time.

Rosenzweig: It gives us a baseline on where we are, but for learning anything about policy it's absolutely useless.

Henderson: But what I'm saying is that you have surveys conducted over periods of time in a number of different countries with similar methodologies so that you have some comparable descriptive data. And that is worth while. But you can never, from those kinds of data, deal with the issue of causation. They give you patterns and so they're useful as a first step. But I think, unfortunately, that we've got to bite the bullet and gather data on households and individuals over time with economists, nutritionists and physicians working together. Perhaps at the same time we should be making interventions so that we have some experimental results. There is a precedent for conducting, on a global basis, comparable surveys in different countries and over time. However, I believe they could be far less elaborate than the World Fertility Survey.

Rosenzweig: That's where we disagree. The World Fertility Survey was one of the simplest surveys ever done. It had no information on income. I mean you could learn nothing form that survey about what was determining fertility except to correlate it with the education of the mother. What you need is a survey that gets good information on incomes and that's very hard to do. You need to get good information on illness and food consumption. I don't see, without that kind of information, that you can draw better inferences than the studies I referred to. When I state these relationships we all recognize that there is a lot that can be wrong. But just collecting health and education information is not going to answer the question. The World Bank has started. They're collecting living standards measurements in a number of countries. They are longitudinal. Each year they keep expanding it because each year we find some inadequacies in the data.

Henderson: I don't think it's insolvable. I think the trick is to try to get surveys which can be done without years of work, enormous investment of time, energy and effort, and that take forever to analyze. One should strive for fairly simple survey barometers of measurement, which can be done in a fairly short period of time. And I think that is do-able.

Rosenzweig: If you just want to measure how many people are ill, by a good definition of illness, it is do-able. It would be just like the World Fertility Survey. You do want to know what the world is like. That's the surveillance. And you could get regional delineations too. There's no question that would be useful. But from a scientific perspective, we want to learn what interventions are useful. To determine relationships between agricultural development and health, they would be totally inadequate. And that's the problem. There is this trade-off between spreading resources thinly across many areas and getting baseline information that's reliable but that takes many years to design, implement, and analyze.

Calloway: I do hope that when you do publish the information, or otherwise get it out, that you will discriminate very carefully between whether you're talking about food disappearance, whether you're talking about food that was actually measured that was going into mouths, or food that was reported to someone else as having been consumed. There's almost no good information on individual food intake anywhere. Don't be too swayed by the information about male food intake. Men are the least reliable reporters of food consumption. They eat a lot away from home. They don't always want to say where they ate if they've spent household money and they seldom report alcoholic beverage consumption accurately.

Rosenzweig: I don't want to be as nihilistic as you. I just want to emphasize what I see, as the major barrier to the scientific understanding of developmental issues, which is the database problem. It has become a little bit clearer to me in this meeting that there are lots of different kinds of databases. We discussed the need for surveillance. Surveillance is a different kind of thing than the kinds of surveys that I'm used to working on. And there is a third kind, which is interventionist experiments, that one can learn about, and maybe ethnographic surveys. But for sure, in the area of health and development, we need some institutions that facilitate the working together of economists, nutritionists, and health specialists to design and implement surveys. I'm not sure of what institutions they should be. But unless this happens, we're not going to make terribly much progress in the near future.

15

Population and Development

Bruce F. Johnston

Johnston: Mark Rosenzweig has summarized what might be termed the "revisionist" perspective on demographic change:

"The observed mix of the large families and the low levels of health, nutrition, and schooling, are symptoms, not causes, of the lack of economic development. Governments and international development agencies should, therefore, focus on promoting or removing impediments to economic development and not on families' decisions about their size."

I find it absolutely mind-boggling how very intelligent people can so often take either/or the view that we're concerned with development or we're concerned with family planning. I feel strongly about these issues, strongly enough to overcome the fact that I'm venturing into areas of health and population that are outside whatever area of expertise I might claim. For many years I've been particularly concerned with the problems of "late developing countries," - a bit of jargon to describe countries that still have an economic structure that is dominated by agriculture, where for example some 50, 60, 70, 80% of their population and labor force continue to depend primarily on agriculture for employment and income.

The demographic transition is characterized by two phases: the first phase is when mortality rates decline and fertility remains at a more or less traditional level. The result is a period of rapid growth. But for the countries in Europe, and even Japan, that rapid population growth was in the range of 1.0 to 1.5 percent per year. In Taiwan, by the 1920s population growth was running at an annual rate of 2.5

percent. By the 1930s even the rate of growth of the labor force had reached 2.5 percent. In contemporary late developing countries--of which there are some 50-60 at the present time--rapid population growth often means 3-4% a year. This means--just a matter of simple arithmetic--that the time required to transform an overwhelmingly agricultural economy is very much extended. This is epitomized by some projections done several years ago for Kenya. Even at the end of a 55 year period, in 2024, some 65% of Kenya's labor force will still depend primarily on agriculture. And that is using pretty optimistic assumptions about the growth of non-farm employment. The projections suggest that over this 55 year period there will be a 4-fold increase in the labor force--in spite of an assumed 16-fold increase in the urban population.

When I was last working in Kenya for an extended period, back in the mid-1970s, I was reintroduced to the child survival hypothesis. When I first encountered that hypothesis, I reacted very negatively, because the early statements of it were overly optimistic and, equally important, they seemed to treat it as something automatic. Carl Taylor himself has always stressed the potential significance of reaching a threshold in reduction of infant and rapid decline in birthrates. The relationship is strengthened if it is associated with programmatic interventions, that make an effective attempt to speed up the awareness of parents of this drastically changed mortality situation and its implications for a desired size of family.

Now the determinants of fertility are obviously extremely complex. I know that I don't begin to really understand the complex determinants of changes in fertility, and I'm very suspicious of anybody who claims to know. But there are a few facts that we do know about the demographic transition. One is that when birth rates do begin to decline, when a country enters this second phase of the demographic transition the decline in fertility often occurs very rapidly. Cleland and Wilson (1987) assert that the explanation for this decline must lie in social or psychological developments, that are capable of rapid transformation. The changes in factors such as income and relative prices emphasized by the so-called demand side theories of fertility just don't occur that rapidly. One factor often emphasized by economists such as Gary Becker is a change in parents' attitudes from an emphasis on the quantity of children to an emphasis on the quality of children. That's an ideational change that could occur very quickly. I'm at a loss as to how we would measure whether such a change had occurred. I'm even more at a loss as to how we might design an informational campaign that would accelerate such a change so that the thinking of parents is in terms of quality rather than quantity.

There are a few other things that seem to be well established. One that Mark mentioned, is the great importance of female education. And that, of course, is relevant to an ideational change that could take place quite rapidly. Fortunately most developing countries, including most in tropical Africa, have in fact one area in which they have made very significant progress. That is expanding the coverage of their educational systems - and the expansion of coverage of females has been more rapid than males so there has been at least some reduction in most developing countries in the discrimination against females.

But the most significant factor in bringing about change in attitudes to family size appears to be parents' perception of the supply of surviving children. In a traditional society parents by and large, do not regard the number of children they're going to have as being a result of conscious decisions. In Kenya as late as the mid-1970s, in spite of the dramatic reduction in infant and child mortality that had already taken place, the great majority of Kenyan women did not perceive that there had been any change in the survival prospects in children born by women in 1974 as compared to their mothers a generation before.

A critical threshold is reached when parents perceive this change in survival prospects. This is likely to be accompanied by changes that also affect their ideas about the desired number of children that they'd like to end up with. The threshold that occurs when parents become aware of the drastically changed situation means that for the first time it becomes usual for family size to reflect conscious decisions by parents because they now perceive a situation characterized by an excess supply of children. Before that the thought of having too many children was simply not a salient consideration for them.

The number of interacting variables is very large. I have tried to challenge some very bright econometricians to tackle the problem but without success. But I think that on many of the crucial issues about development strategy we have to be guided by a strategic notion of what is probably right even though we can't adequately back it up with an econometric analysis. The changes in total fertility rates and other vital statistics in low and middle-income countries and in a set of 16 countries that experienced "significant" declines in fertility since 1950-1955 are extremely suggestive (Table 15.1). In particular, the countries with "significant" declines have had outstanding success in lowering infant and child mortality rates.

Warren: I am much more optimistic. Even a few decades ago there was no evidence anywhere that fertility was going to decline drastically. Now, throughout Asia and Latin America it appears that the trend has started down. Africa is the only exception, and it only has about 16 percent of the world population.

Johnston: But it would be demographically impossible for it to turn down in the next 20 years.

Bradley: Bruce, I had understood that there is some evidence in African among higher income groups on some of the countries of declines in total fertility numbers. The better educated and better off people have begun to control their fertility. I assume it will cascade at some rate down the class and the income structure. Am I right in believing that there is beginning to be some evidence that the decline is beginning?

Johnston: Kenya was the first African country to start a family planning program. For some time there has been substantial use of contraception in Nairobi and some in two or three of the other major cities. Obviously, this increase in use of contraception in Nairobi has done little to affect national fertility or the rate of natural increase wouldn't be in the 4 or 4.1 percent range. The other evidence that is more encouraging comes from Zimbabwe and Botswana, where in recent years there appears to be fairly wide-spread acceptance of family planning that may be beginning to have an effect on the rate of natural increase.

Ruttan: Bruce, what about this argument that the African family structure is so different than the Indo-European family structure that one simply cannot expect the same set of forces to operate in the same manner in Africa?

Johnston: That's essentially the position the Caldwells took in their 1988 article. But in a World Bank paper he seems to have backed off

TABLE 15.1 Total Fertility Rate (TFR), Crude Birth Rate (CBR), Crude Death Rate (CDR), Rates of Natural Increase, Infant Mortality Rates and Child Death Rates of Various Groups of Countries and in Individual Countries with Significant Declines in Fertility

Country / Group	Total Fertility Rate 1955-60	Total Fertility Rate 1980-85[a]	Crude Birth Rate 1960	Crude Death Rate 1960	Crude Birth Rate 1985	Crude Death Rate 1985	% Rate of Natural Increase 1985	Infant Mortality Rate 1960	Infant Mortality Rate 1985	Child Mortality Rate 1960	Child Mortality Rate 1985
Average for 35 low-income countries	na	5.9[b]	49	18	43	15	2.8	165	112	28	19
Average for 96 middle-income countries	na	4.3[b]	43	17	32	10	2.2	125	68	23	8
Average for 16 countries with significant declines	5.8	3.2	41	12	24	7	1.8	89	34	12	2
Brazil	6.2	3.8	43	13	29	8	2.1	118	67	19	5
Chile	5.2	2.6	37	12	22	7	1.5	114	22	18	1
China	5.4	2.4	39	14	18	7	1.1	165	35	26	2
Colombia	6.7	3.9	46	14	27	7	2.0	93	48	11	3
Costa Rica	7.1	3.5	47	10	29	4	2.5	71	19	6	d
Cuba	3.8	2.0	32	9	17	5	1.2	66	16	5	d
Jamaica	5.0	3.4	39	10	25	6	1.9	52	20	3	1
Korea, Rep. of	6.0	2.6	43	13	21	6	1.2	78	27	9	2
Malaysia	6.9	3.9	45	16	30	6	2.4	72	28	7	2

Country / Group	Total Fertility Rate 1955-60	Total Fertility Rate 1980-85[a]	Crude Birth Rate 1960	Crude Death Rate 1960	Crude Birth Rate 1985	Crude Death Rate 1985	% Rate of Natural Increase 1985	Infant Mortality Rate 1960	Infant Mortality Rate 1985	Child Mortality Rate 1960	Child Mortality Rate 1985
Mauritius	6.0	2.8	36	8	20	7	1.3	65	25	9	1
Panama	5.9	3.5	41	10	26	5	2.1	68	25	5	1
Sri Lanka	5.4	3.4	36	9	25	6	2.1	71	36	7	2
Taiwan	na	2.8	39	10	21	8	1.3	56	25	8	1
Thailand	6.4	3.5	44	15	26	8	1.8	103	43	13	3
Trinidad and Toboga	5.3	2.9	38	9	25	7	1.8	45	22	2	1
Turkey	6.2	4.0	43	16	30	8	2.2	190	84	50	9

a. Developing countries with an estimate TFR of less than 4.0 for the 1980-85 period with certain omissions: Singapore and Hong Kong as city states, Argentina and Uruguay because their TFRs were only 3.1 and 2.8 in 1950-55, and Puerto Rico and Lebanon because of their special circumstances.

b. The population-weighted average for low-income countries exclude China and India where the 1985 TFRs were 2.3 and 4.5 respectively.

Source: World Development Reports 1978, 1982, 1983, and 1987; Infant and Child Survival; John A. Ross et al.; Family Planning and Child Survival, 100 Developing Countries, Center for Population and Family Health: Columbia University, New York.

a little from that and takes a position that it is going to be more difficult, but not impossible. I think it is dangerous to take a defeatist attitude. But if you have to depend on increases in per capita income as the key factor leading to population decline the situation in Africa is exceedingly bleak.

PART SIX

Perspectives and Reflections

16

Institutional Design

Donald A. Henderson and David E. Bell

Henderson: I have been thinking about the of the tropical diseases that influence labor productivity and land-use patterns. In Africa it is malaria, trypanosomiasis, river blindness, and schistosomiasis. In Latin America it is malaria. There now exist two programs which could serve as building blocks for international research in two of the diseases. The onchocerciasis control program (OCP) has some characteristics of an international center. For the animal disease component of trypanosomiasis, there's ICIPI. In addition, the tropical disease research program of WHO concerns itself with the three other diseases (plus others). This isn't much, but it's a beginning.

It seems to me that the advantages of an international center model are compelling. They can have defined long-run goals, and regular evaluation of progress. Much can be achieved with an appropriate multi-disciplinary staff, supporting infrastructure, and high levels of scientific and technical competence. With multi-country programs and sustained international financing they would be more or less resistant to short-run national government political pressures while still being located in the areas where the problems are greatest. That's a fairly strong case for thinking seriously in terms of such an initiative in spite of the anti-center sentiment in some circles. There has been a very positive experience in agriculture. And I haven't really heard a good argument for why a similar approach wouldn't work in health.

Warren: There was a schistosomiasis center which lasted for 16 years. That was the Rockefeller Foundation's experiment on the island of St. Lucia. It was fine while it lasted.

Bell: That wouldn't quite fit the standard definition of an international center. It was a very important project but it didn't, for example, have the objective of turning out important products to be used in schistosomiasis control around the world. It was aimed as an important set of research issues, and well-designed for that purpose.

Warren: St. Lucia's primary goal was to validate and sort out control strategies. The strategy that came out of St. Lucia is the key strategy that's being used today in the control of schistosomiasis.

Bell: I have several points that don't hang together very well. The first is that, lurking in the background of this meeting was the question, if I understood it rightly, posed by Bob Herdt and Vern, whether there might usefully be a somewhat more organized conference. My own response to that question, based on this meeting, is yes, indeed. This was a stimulating, productive interchange. It was pretty spotty. There weren't any prepared papers in a strict sense. Nobody tried to give us a framework of interrelationships between health and agriculture that we could shoot at. But it seems to me that enough things came up that are of significance so that it would be feasible and desirable to think about a larger and better-organized conference. What would it be about? I would think possibly it might be aimed at encouraging health and agricultural researchers to talk to each other and to develop ideas about institutional relationships and research agendas. In other words, it would not simply be a conference to explore the subject further, but to try to aim at the kind of combined research that many of us have talked about during this meeting. (Ruttan, 1994)

We found ourselves talking a lot about nutrition--nutrition, health and agriculture. We found ourselves talking a lot about specific illnesses and diseases--schistosomiasis, malaria and Dave Bradley's list. I think that we suffered from the fact that nobody had been asked before we got here to prepare what I called yesterday an index of impact, which would include some attempt to bring together some numbers about severity. A case of malaria isn't equal to a case of onchocercias. Some of these, like onchocercias affect quite small areas on the earth's surface, and others, like malaria, affect very large areas. How do you bring these together in some kind of common scale of importance?

The third area that Vern outlined before we got here and it indeed turned out to be a rich subject for varied discussion, is the environmental impact of agricultural development - which probably converts into something broader than environment because a lot of what we were talking about were social responses, not strictly speaking the impact on natural resources. Did people move, or didn't they? What difference does it make? How do people deal with orphans? We're not talking only about how agriculture affects physical resources, but how it affects the social fabric. And not simply in disease terms but in broader terms.

Finally, we have identified institutional questions of much interest. One is the desirable interplay between health and agriculture research. I include nutrition and population in a broad concept of health as the Commission did in its recent report. But we've also turned up other institutional questions. One is Mike Osterholm's comments about the globalization of food supply. The institutional question that raises is, "How are we going to globalize the creation and enforcement of standards of cleanliness, of chemical impact?" It isn't enough that Gene Allen is careful about how many tomatoes from Mexico he eats. We need something that's a bit more orderly and institutional, it seems to me. And it needs to be double-ended. The people you have here at this university in your department of Agricultural Economics, and your students from Mexico, who are going to be going home and working in Mexico, are the logical collaborators to address a question dealing with the health effects of tomatoes grown in Mexico and eaten in the United States, or grown in Mexico and eaten in Mexico. We should not impose U.S. standards on the rest of the world through our trade rules. We are dealing, in some sense, with the development of institutional relationships among scientists and policy-oriented people in countries around the world. This is one institutional corollary of some of the things we've been talking about.

To wind up these wandering remarks, it seems to me that we do have, in the subject of health and agriculture, not the original narrow agenda that Vern suggested -- but the subject of health and agriculture viewed as an interrelated system. We have a very lively and stimulating area. It opened up a lot of novel and interesting and unfinished conversations, whether it's a conversation between the epidemiologist and the economist, or whether it's a conversation about statistical standards for surveys between Mark and some of the physicians. All of these are simply the opening stages of interesting and important conversations and discussions attempt to build better understanding of the objective of assuring healthier people in a richer and more sustainable agricultural society in the world.

Henderson: After Dave's more mature remarks, I'm not sure how much more I can add. I'd like first of all to go back to remind you again of the concern that some of us have, not an unrealistic concern, that the health area has not received very much attention. And I would like to refer back again to the possibility that wild cards could again occur. We went through the 1960s and 1970s in the health field proclaiming that we had defeated infectious diseases. And I can extract pieces from the Surgeon General's comments on various dedications and what-have-you that chronic diseases were now what we should be concerned about.

Now we have a disease called AIDS, which really upsets an entire balance of factors, with regard to population, economic factors, political factors. It's not the only new disease we've seen in recent years which has emerged. Lyme Disease, I would point out, is a fairly recent phenomenon. The fact that we now have Hemorrhagic Fever in the Caribbean and extending through Latin America, and becoming a major problem in Southeast Asia, is yet another new phenomenon. Very little attention has been given to this yet, but we certainly have the potential for major epidemics. We have Legionnaires Disease. Can we expect others?

It is useful to recall the great experiment of introducing Myxomatosis into Australia to take care of the rabbit population. There was a mortality rate of close to 99 percent for the first introduction of that virus into the rabbit population of Australia. Were this a human disease, it would be a horror tale. We don't like to think about such a scenario, but I don't think that we can rule out the possibility of other particularly virulent viral diseases giving us as much or more trouble than AIDS. At this time, our concern is that we don't have the capacity to address these problems. This certainly impacts all of us in the health field, and it impacts agriculture. We have lead poisoning, for example, which is an emerging problem of far greater significance than we had ever appreciated, with permanent damage done to children at a very young age inducing mental retardation at very much lower levels than we had ever believed could cause problems. Lead is an inert element which does not readily combine and disappear in an innocuous form. With the production of lead still high, it is becoming a real problem in urban areas around the world. It is mostly being produced in the United States. In the United States the production is actually going up. It's being exported. I think this is going to be, for Eastern Europe as well as for the rest of the world, a very major problem. I only cite it as a problem which we had not talked about but that looks rather worrisome to all of us. It could have profound consequences to the agricultural work force.

I'm glad that Michael referred to the globalization of food supplies. This is a new phenomenon, where I think uniquely health and agriculture do come together. They must come together in a way that they have not done before. In passing, I would like to cite an interesting experiment that we conducted some years ago. We were having a lot of hepatitis in the western part of the United States. It had a lot to do with lettuce from California. We began to contemplate the possibility that perhaps there were people defecating on the irrigation canal banks and hepatitis was somehow or other contaminating the lettuce. Then we made an interesting observation. The observation was that the size of the hepatitis virus was actually smaller than the pore size of the roots of the lettuce. So we took celery and we grew it in a solution of polio virus, as the polio and hepatitis viruses are a similar size, and, guess what? We were able to isolate the polio virus from the leaves of the celery. Further studies were not carried out. This could represent yet one more dimension to the food-health problem.

There are other important challenges based on what we know about methods for inspection of products, whether they be canned or fresh. It seems unlikely that we're going to devise mechanisms by which we're going to be able to screen or monitor these various foodstuffs coming into the country. It has the implications of protectionism and all the other problems if this gets out of hand. This is a really major problem area which demands being addressed. Very interestingly, our only real screening method for problems is basically a body count. As soon as we've got enough sick people or dead people, we then go back and look for what's gone wrong.

We don't really have a very sensitive mechanism for picking up illnesses and pursuing them. I don't know what Mike's staff is, but I'm sure it's quite a small number trying to deal with a major set of problems and Jim's staff in California was minute for that enormous state. I would submit that one of the problems that we have in common and approach separately is the objective of what we're all about, and that is - healthy, productive people. In agriculture, you're measuring progress in terms of additional bushels per acre. We're looking at how many sick people we're treating in hospitals, how many vaccinations were performed. But in fact, for neither of us is this the object. Our ultimate objective is healthy people, and not foodstuffs or vaccinations. So it comes back again to what I feel is very critical for the future. We have got to devise better means for monitoring the condition and the health of people--growth patterns, illness patterns, what have you --and to pay a great deal more attention to the question of how we produce healthier and more productive people. This is where we come together around a common objective.

17

Implications and Priorities

Vernon W. Ruttan

Ruttan: Evidence on the question raised at the beginning of the consultation - does health represent a serious constraint on agricultural development? - is at best ambiguous. Scattered data from countries such as India, Indonesia and Ivory Coast indicate loss of days worked due to sickness in the 5-15 percent range. In the USSR and Poland, substantial numbers of days of work are lost due to respiratory disease associated with atmospheric pollution.

There have been major "plagues" in the past that resulted in mortality levels sufficient to seriously impinge on food supply. In the fifteenth century following the Spanish conquest, the Amer-Indian population in the basin of Mexico declined by something like 90 percent. Most of the decline was due to a series of epidemics - smallpox, measles, typhus and plague. Famine, associated with the high dependency to working adult ratio, probably accounted for 10-15 percent of the population loss.

The population loss from most historical plagues in Europe and Asia were concentrated in the younger and oldest age groups rather than among the adult population of working age. Many adults had survived earlier attacks and had acquired some degree of immunity. The incidence of death from the European and Asian diseases introduced into the Americas was spread more evenly across the age distribution because everyone was equally susceptible. The AIDS plague is unique in that it is killing people who would be at their most productive age. The result will be a rise in the dependency ratio -- the ratio of the old and young relative to workers in the more productive age groups. There are important questions that have not yet been sorted out in the relationships between AIDS and other diseases. One apparent consequence of AIDS in East Africa is a rise in tuberculosis.

The World Health Organization has an active program of cooperation with Africa and other high incidence AIDS countries in estimating HIV infection and AIDS incidences. A further step should be an intensified to model the direct and interaction effects of the simultaneous incidence of HIV infection and tropical parasitic and viral diseases on morbidity and on mortality.

Specific Issues

Let me now list some of the more specific research implications that emerged from the consultation.

The capacity to design systems of health delivery that are capable of reducing the incidence of illness continues to elude health policy and planning agencies in both developed and developing countries. The systems that are in place in most countries can be more accurately described as sickness recovery systems rather than health systems. A major deficiency is the lack of a system for providing families and individuals with the knowledge needed to achieve better health with less reliance on the health care system. The point was made several times during the consultation that almost all countries have been able to design reasonably effective agricultural extension or technology transfer systems to provide farm people with the knowledge about resources and technology needed to achieve higher levels of productivity. We have yet to design an effective system to provide families and individuals with the knowledge in the area of human biology, nutrition, and health practice that will enable them to lead more healthy lives.

The residuals produced as a by-product of industrial and agricultural production have become an ever increasingly important source of illness in a number of countries and regions. The most serious impacts are occurring in the centrally planned economies of Eastern Europe, the USSR and China. Levels of atmospheric, water and soil pollution have resulted in higher mortality rates and reductions in life expectancy. The effects are evident in the form of congenital malformation, pulmonary malfunction and excessive heavy metals in soils and in crops grown on contaminated soils. Many of the health effects of agricultural and industrial intensification are due to inadequate investment in the technology needed to control or manage contaminants. Rapid industrial growth in poor countries, in which investment resources are severely limited, will continue to be accompanied by underinvestment in the technology needed to limit the release of contaminants. The situation that exists in Eastern Europe presents a vision of the future for many newly industrialized countries unless better technology can be made available and more effective management of environmental spillover effects can be implemented.

Lack of location specific or site-specific research capacity represents a major constraint on the capacity of health systems in most developing countries. It is no longer possible to maintain the position that health related research results can simply be transferred from developed country research laboratories or pharmaceutical countries to practice in developing countries. Local capacity is needed for the identification and analyses of the sources of health problems. It is also needed for the analysis, design and testing of health delivery systems. The international donor community has been much slower in supporting the development of health research systems than agricultural research systems in the tropics. For example, there is now in place a network of more than a dozen international agricultural research centers (IARC's), sponsored by the Consultative Group on International Agricultural Research that play an important role in backstopping national agricultural research efforts. The only comparable international supported center in the field of health is the Diarrheal research Center in Bangladesh. Furthermore, the capacity to conduct research on tropical infectious and parasite diseases that was supported by the former colonial countries - United Kingdom, France, Netherlands and Belgium - has been allowed to atrophy.

High birth rates are both a consequence and a cause of poor health. The demographic transition - from high to low birth rates - has in the past usually followed a rise in child survival rates. This suggests that improvements in health, particularly of mothers and children, is a prerequisite for decline in population growth rates. But high population growth rates, particularly in areas of high population density, are often associated with dietary deficiencies that contribute to poor health and high infant mortality rates.

The issue of how to achieve high levels of health and low birth rates at low cost in poor societies remains an unresolved issue. Several very low income countries have achieve relatively high levels of health - as measured by low infant mortality rates and high life expectancy rates -but often a high cost relative to per-capita income. Other societies that have achieved relatively high incomes continue to exhibit relatively high infant mortality rates and only moderately high life expectancy levels.

More effective bridges must be built, both in research and in practice between the agricultural and health communities. At present these two "tribes", along with veterinary medicine and public health, occupy separate and often mutually hostile "island empires". But solutions to the problem of sustainable growth in agricultural production and improvement in the health of rural people and the consumers of agricultural commodities requires that each of these communities establish bridgeheads in the other's territory. Multi-purpose water resource development projects have contributed to the spread of onchocerciasis. Successful efforts to control the black fly have reopened productive lands to cultivation. The introduction of improved cultivators and fertilization practices have helped make the productivity growth sustainable. But examples of effective collaboration, either in research or in project development, are difficult to come by.

Some Generic Issues

A set of four generic issues ran through the earlier consultations on agriculture and the environment, as well as this consultation.

The first is that many of the problems that we have discussed are transnational in scope. This means that many of the institutions that will be needed to enable societies to respond to the constraints on sustainable increases in agricultural production will have to be transnational or international. We can no longer get by with slogans such as - "Think globally and act locally". We will have to institutionalize the capacity to respond to scientific, technical, resource, environmental and health constraints. In the area of health, for example, it seems clear that almost every source of illness or poor health that exists somewhere - whether the source is an infectious organism or environmental change - will exist everywhere else. This statement may be an exaggeration - but it is only a slight exaggeration.

A second is our limited capacity to design the institutional infrastructure that will be needed to sustain the required rates of growth in agricultural production as we move through the first decades of the next century. We need to build institutional infrastructures that facilitate more effective collaboration among engineers, agronomists and health scientists - to deal with issues of production, environmental change, and the health of food producers and consumers. The social science disciplines and related professions (law, management, social service) have not demonstrated great capacity in the area of institutional design. Plant breeders have been much more effective. They don't just analyze the sources of yield differences - they utilize the agronomic and genetic knowledge that is obtained from their analyses to design improved cultivators - plants and animals that are responsive to management and that are resistant to the assaults of nature. In the social sciences, once we complete our analysis, we feel that our job has been finished. We tend to stop at the level of analysis. We only rarely bring the knowledge we have acquired to bear on institutional design.

The third is the inadequacy of our capacity to monitor changes in the sources of productivity change, environmental change, and the insults to health. We know very little about either the levels or the trajectories. We talk about soil erosion, but we don't have the monitoring capacity to know the extent to which it is weakening our capacity to produce. We are fighting a defensive battle against the health effects of the contamination of our food supply rather than anticipating the sources. One of the puzzling aspects of the data that has become available so far is that the health effects of increased use of fertilizer is less than expected in spite of high levels of nitrate in surface and ground water. Neither the developed or developing countries have in place adequate surveillance systems for disease. Our knowledge of the sources of morbidity is exceedingly weak.

18

Agricultural and Health Research: Bridging the Island Empires

David E. Bell and Vernon W. Ruttan

The purpose of this chapter is to outline a vision of the health and agricultural research systems that must be completed, or put in place, in order to generate the new knowledge and the new technology necessary to meet the demands for food security and health maintenance in the early decades of the twenty-first century. In developing this vision we draw heavily on a series of consultations, held during 1988-90 and an international conference involving leading agricultural, health and environmental scientists held during November of 1991.[1]

Two perspectives emerged from the consultation and the conference. One is that the battle to achieve sustainable growth in agricultural production must be fought out along a broad multi-disciplinary front. Poverty undermines health and degrades the environment. Environmental problems such as soil erosion, water logging and salinity, and fertilizer and pesticide residues link the agricultural agenda with issues such as malaria and schistosomiasis control, sanitation, and water and food quality on the health agenda. Environmental changes underway at the global level, such as acid rain, ozone depletion and climate change will require changes in food production

---

[1] The dialogues and recommendations from the initial three consultations, held under the auspices of the "Twenty-First Century Project," with support from the Rockefeller Foundation have been reported in three University of Minnesota Department of Agricultural and Applied Economics Staff Papers (Ruttan, 1989; Ruttan, 1990a; Ruttan, 1990b). A revised version of the second consultation report has been published by Westview Press (Ruttan, 1992). The results from the conference, held at the Rockefeller Foundation Bellagio Conference Center are reported in Vernon W. Ruttan (1994).

and health practices at the producer and community level. Effective bridges must be built between the "island empires" of the agricultural, environmental and health sciences.

A second perspective is the central role of decisions by the family and the community and in achieving growth of agricultural production, enhancement of the resource base, and improvements in health. This means that much more effective organizational and institutional linkages must be built between the suppliers of knowledge and technology and the users. It also means that the institutions must be designed to place the users in a stronger role relative to the suppliers.

During the discussions at the 1991 Bellagio conference a vision of the institutional infrastructure that will be needed to supply the knowledge and technology in the areas of agricultural production, resource management, and health began to take shape. In this chapter we draw on the papers and discussion at the Bellagio conference and at the three earlier consultations to outline our vision of the structure of global agricultural, health and environment research systems. We are under no illusion that the process of evolving an effective global research system that will be capable of bridging the island empires of agriculture, environment and health will be easy.

Agricultural Research

This vision is strongly influenced by the experience of attempts, beginning in the late 50s to establish a global agricultural research system (Ruttan, 1986; Baum, 1986). For the architects of the post-World War II set of global institutions meeting world food needs and the reduction of poverty in rural areas were essential elements in their vision of a world community that could ensure all people of freedom from hunger.

In the immediate post-war years much of the burden fell on the United Nations Food and Agriculture Organization (UN/FAO). But John Boyd Orr, the first Director General of the FAO, burdened with the memory of the agricultural surpluses of the 1930's was highly critical of the view that knowledge and technology represented a serious constraint on agricultural production capacity, "No research was needed to find out that half the people in the world lacked sufficient food for health, or that with modern engineering and agricultural science the world food supply could easily be increased to meet human needs." (Boyd-Orr, 1966:160) In the first two post-war decades assistance for agricultural development in the poor countries was conducted largely in a technology transfer and community development mode. By the late 1950s, it was becoming apparent, however, that the gains in production from simple technology transfer had largely played themselves out.

The inadequacy of policies based on the technology transfer or extension model led, in the early 1960s, to a re-examination of the assumption about the availability of a body of agricultural technology that could be readily diffused from high agricultural productivity to low productivity countries or regions. The result was the emergence of a new perspective that agricultural technology, particularly yield enhancing biological technology, is highly "location specific." Evidence was also accumulated to the effect that only limited productivity gains could be achieved by the reallocation or more efficient use of the resources available to peasant producers in poor countries.

The new vision that emerged as a guide to the sources of growth in agricultural production was the product of both experience with the improvement in agricultural technology and a reinterpretation of the role of peasant producers in the process of agricultural development.

In the early post-war development literature peasant producers had been viewed as obstacles to agricultural development. They were viewed as bound by custom and tradition and resistant to change. In an iconoclastic work published in 1964 Theodore W. Schultz advanced a "poor but efficient" view of peasant producers. They were viewed as making effective use of the resources available to them. But they lived in societies in which productivity enhancing "high pay-off" inputs were not available to them.

Schultz, drawing on the experience of the Rockefeller Foundation program in Mexico and case studies by anthropologists and agricultural economists elsewhere, identified three "high pay-off" investments needed to enhance the productivity of peasant producers. These were: (a) the capacity of the agricultural research system to generate locally relevant knowledge and technology; (b) the capacity of the industrial sector to develop, produce and market new inputs which embodied the knowledge and technology generated by research; and (c) the schooling of rural people to enable them to make effective use of the new knowledge and technology.

These insights, from experience and analysis, shaped the response to the food crises of the 1960s and 1970s. The immediate response was the transfer of large resources, including food aid, to the food deficit countries. The longer term response was the mobilization of resources to develop a system of international agricultural research institutes and to strengthen national agricultural research systems.

The leadership of the Consultative Group is now centered at the World Bank, which provides a chairperson and a secretariat. Each institute or center is an independent corporate identity governed by its own board of trustees. The CGIAR established a Technical Advisory Committee (TAC) with its secretariat located at FAO in Rome, to provide technical oversight of the research institutes and to advise the CGIAR on priorities and resource allocation among centers. The TAC has been charged with the responsibility of organizing comprehensive reviews of the programs of the centers, of evaluating new initiatives, and of overseeing coordination among centers in common program areas such as cropping systems research.

By the early 1990s the system had expanded from an initial 4 to 18 centers. The initial centers focused their research on the major food crops grown in developing countries -rice, wheat, maize, potatoes and cassava. These were joined in the 1970s by centers focusing on livestock production, animal disease and genetic resources, on arid and semiarid areas, food policy, and the capacity of the national research system. At the beginning a commodity orientation in research and development was adhered to in an effort to assure that the limited resources available to the system would not be dissipated in unfocused research efforts.

As the new seed-fertilizer technology generated at the CGIAR centers, particularly for rice and wheat, began to come onstream some donors assumed that the CG centers could bypass the more difficult and often frustrating efforts to strengthen national agricultural research systems. But experience in the 1960s and the 1970s confirmed the judgement of those who had participated in the organization of the international centers that strong national research centers were essential if the prototype technology that might be developed at the international centers was to be broadly transferred, adopted and made available to producers.

The location-specific nature of biological technology meant that the prototype technologies developed at the international centers could become available to producers in the wide range of agro-climate regions and social and economic environments in which the commodities were being produced only if the capacity to modify, adapt and reinvent the technology was available. It became clear that the challenge of constructing a global agricultural research system capable of sustaining growth in agricultural production required the development of research capacity for each commodity of economic significance in each agro-climactic region. One response by the CGIAR donor community was the establishment of a new Center, the International Service for National Agricultural Research (ISNAR) to provide analytical and technical assistance to national agricultural research systems in strengthening their organization and management. Another response was, particularly during the 1970s, substantially expanded support for national agricultural research systems.

During 1990-92 five new centers were added to the CG system thus increasing the number of centers from 13 to 18. In 1990 the International Irrigation Management Institute (IIMI), the International Center for Research on Agro-Forestry (ICRAF); and the International Network for the Improvements of Banana and Plantain (INIBAP) were brought into the CG system. In 1992 the International Center for Living Aquatic Research Management (ICLARM) was added to the system.

Expansion of the international agricultural research system was not accompanied by a comparable expansion of the resources available to the system. Support to the system since 1991 has actually declined in real terms producing a "quiet crisis in the system". The crisis has not been only financial. A number of the CGIAR centers are experiencing the difficulties associated with organizational maturity. There is a natural "life cycle" sequence in the history of research organizations and research programs (Ruttan, 1982:132). When they are initially organized they tend to attract vigorous and creative individuals. As these individuals interact across disciplines and problem areas the organization often experiences a period of great productivity. As the research organization matures, however, there is often a tendency for the research program to settle into "filling in the gaps" in knowledge and technology rather than achieving creative solutions to scientific and technical problems. Since the early 1980s a number of the managers of several of the CGIAR institutes have been forced to struggle, during a period of budget stringency, with the problem of how to revitalize a mature research organization.

Efforts to strengthen national research institutes have also been only partially successful. The 1970s witnessed a remarkable expansion of agricultural research capacity in a number of developing countries. The national research systems in India, Brazil, Malaysia and a number of other developing countries began to achieve world class status in their capacity to make advances in knowledge and technology available to their farmers. A number of other countries, such as the Philippines,

Colombia, Kenya, and Thailand achieved substantial capacity to conduct research on their major agricultural commodities. By the mid-1980s the buffeting of a global debt crisis and the weakening of commodity markets had the effect of dampening commitment by a number of aid agencies and national governments to the strengthening of agricultural research. In Africa many national agriculture research systems that have received generous external support even during the 1980s have failed to become productive sources of knowledge and technology (Eicher, 1994).

The role of technical support for farm decision making by farmers and the capacity to supply to producers the technical inputs in which the new technology is embodied has been a continuing area of controversy. In general the developing countries have been relatively extension intensive. The ratio of extension workers to agricultural product has been much higher in developing countries than developed countries (Judd, Boyce and Evenson, 1987). Weak linkages between research and extension and between extension and farmers have represented a serious constraint on the diffusion of new technology (Tendler, 1994). During the late 1970s and early 1980s the World Bank devoted very substantial resources to the support of an intensive "training and visit" (T&V) system of delivering information about practices and technology to farmers. The system involved a highly regimented schedule in which the field level worker is involved one day each week in intensive training about the information that he or she must convey to farmers (Benor and Harrison, 1977). In retrospect it appears the system erred in placing the extension worker rather than the farmer, or the farm family at the center of the technology adoption process.

A second constraint on the effectiveness of the transfer of agricultural practices and technology to producers has been the weakness of the private sector as a source of both the supply and delivery of knowledge and technology (Evenson, Evenson and Putnam, 1987; Pray, 1987). The emergence of more liberal economic policies since the early 1980s in a number of developing countries is, however, leading to rather rapid growth of private sector suppliers of agricultural technology and to increased research by the suppliers.

The global agricultural support system is still incomplete. Their deficiencies continue to deprive farm families the support that they need to meet even current food consumption and income needs (Eicher, 1994; Turner and Benjamin, 1994, Tendler, 1994). Yet the vision of the agricultural support system that will be needed to sustain growth in agricultural production is reasonably clear. During the past several decades implementations of the vision have been less than adequate in some developed countries and in all but a few developing countries. With the ending of the cold war it may now be possible to extend the vision to farm families in many of the formerly centrally planned economies. One important step will be to place farm families and the farm enterprise in those societies at the center of the agricultural production process. Another important step will be to link the agricultural research systems in the formerly centrally planned economies with the emerging global agricultural research system.

Health Research

The gains in health status that can be achieved by even a poor society that devotes significant resources in support of an effective national health policy has been outlined by Godfrey Gunatilleke (1994). Sri Lanka has achieved health indicators - a life expectancy of around 70 years and infant mortality below 20 per 1,000 live births - comparable to the levels achieved by many societies that are much more affluent. But a vision of the global health research system needed to sustain national health policy has emerged more slowly than the vision of a global agricultural system. Only within the last decade has the health research community begun to articulate the form that such a system might take.

For most of the century - since the time of Koch and Pasteur - health research has been thought of principally as laboratory-based biomedical research, seeking "silver bullets" against specific infections or diseases - new vaccines, new drugs, new surgical techniques. This focus, plus the remarkable improvements in health in recent decades, led to the misperception that all the new knowledge and new technology needed to protect families and communities around the world from debilitation and illness could be generated in the universities, research institutes, and pharmaceutical company laboratories of the industrialized countries. This limited conception was clearly wrong and has been changing rapidly. Three gains in perception are especially important.

The first is the recognition that health technologies, to be useful, must be applied in particular social settings. Achieving health improvements requires not only technology but policies, organizations, and processes that are adapted to the varied economic, social, cultural, and historical circumstances among and within countries. Even vaccines, the simplest of technologies, cannot be applied in Lagos by the same means they are in Liverpool.

An effective health research system, capable of conducting the essential national health research (Lucas, 1994), needs epidemiologists, economists, management specialists, and other social and policy analysts in addition to biomedical scientists. Such skills are scarce in industrialized countries. They are grossly deficient in developing countries. But they are essential to identify the precise nature of health problems in different national and local settings, and to design, test, and apply appropriate solutions.

A second gain in perception is the recognition that the principal actors in achieving improvements in health are individuals and families, especially mothers. Preventing illnesses and promoting health depends first and most of all on "maternal technology" - the ability to implement basic knowledge about nutrition, cleanliness, home remedies, and when and how to call on health professionals (Mata, 1988). An effective health research system, therefore, must be organized not simply to serve physicians but to support the flow of health knowledge and technology to families and communities - and to provide for the reverse flow of information from families and communities to researchers about the actual nature of health problems and how they are changing. Such a conception of linking researchers directly to primary actors is customary in agriculture, where research results have long been aimed at farmers as decision makers. But it is a recent conception in health even in industrialized countries.

A third gain in perception is the recognition that the world's health research efforts are overwhelmingly concentrated in industrialized countries, seeking technologies to address the diseases of the more affluent societies. Only about five percent of global health research financing is directed to the major diseases and health problems of the less developed countries, where more than 90 percent of the world's burden of preventable deaths occur (Commission on Health Research for Development, 1990). An effective global health research system must address this huge imbalance, and provide for a large increase in the resources devoted to the health problems of the developing countries. Combining these three perceptions with the traditional power of biomedical research, one can begin to perceive, dimly, the shape of a global health research system and how to move toward it.

Such a system - in health just as in agriculture - will need to be based solidly on national research systems, capable of supporting decision makers as they identify and confront health problems. A national health research system requires first of all skills to measure the patterns and determinants of disease, disability, and death, and to monitor changes in health status over time. It requires also skills to design, test, and evaluate means for applying improved health technologies in local environments, and for making research results available to those who need to use them, from national policy makers to local families. Every nation needs the capacity to conduct such a country specific research to guide its health activities, and the establishment of such capacity should clearly be given top priority.

Beyond the capacity for essential country specific research, health scientists in every country will wish to join, as and when they can, in the international effort to advance the world's frontiers of knowledge on the social and biological pathologies of ill-health and disability, and on new technologies to overcome them. In poor countries, the conditions for world-class science are difficult to establish. Nevertheless, a significant number of developing countries - to name just a few, Thailand, India, Egypt, Mexico, Brazil - are beginning to have the capacity to make significant contributions to world knowledge in the health field.

Thus, national health research systems need to begin with the capacity to guide national health activities, and to go on, as conditions permit, to participate in global frontier research. In most developing countries, there are only rudimentary health research capabilities at present. It is urgent for developing countries, and for the international health assistance community, to commit themselves to building steadily stronger national health research systems. Such systems will need to start small, and to focus initially on the most pressing health problems. But they should be designed with a view to dynamic change over time as financial and personnel resources grow, and as health problems change with the demographic and epidemiologic transitions through which the developing countries will pass over the coming decades.

Thinking about how to achieve an effective global health research system thus begins with the development of strong national systems. But national systems must not be thought of as separate, free-standing entities. On the contrary, it is essential that they be linked together by strong international ties, and draw from the common, growing pool of world-wide health knowledge, with each country adapting advances in health science to its own specific circumstances.

Moreover, it would be a mistake to think of a global system as centered in the industrialized countries, with all scientific advances pioneered there and rippling outward to the developing world. We have already seen major health improvements developed in the Third World, as ambulatory therapy for tuberculosis was pioneered in Madras, and oral rehydration therapy for diarrhea in Dhaka. As the amount and quality of developing country research steadily rise, a global research system will increasingly be multi-centric - one in which the flows of ideas and new knowledge move in all directions along networks of information and collaboration encompassing scientists from many countries rich and poor alike.

Thus the guidelines for moving toward a global health research system include (1) the development as rapidly as feasible of strong national systems, especially in developing countries where they are currently very weak, and (2) the rapid evolution of international collaborative mechanisms and arrangements. There is much work here for years to come. Two aspects of this overall vision deserve special attention and illumination.

The first is the necessity for building direct relationships between the national health research system and action for health at the community and family level. Dan C.O. Kaseje has described the elements of a community based health system in Kenya that he helped design and implement that relies directly on the actions of individual families and communities (Kaseje, 1994). The model views the mother as the key health provider. It builds on the strong motivation to carry out her tasks resulting from concern about the current and future well-being of her children and family. Kaseje summarizes the concept behind the "Harambee" model:

This model recognizes the strengths and resources of the community; seeks to facilitate and enhance these strengths; recognizes that communities have always been responsible for their own health, even without the intervention of health professionals; that the mother is the most important and knowledgeable health provider. The mother is not, however, left without resources to carry out her responsibilities. She is reinforced with a strong program of health education, the availability of appropriate technology and materials, and support from NGO and official health programs. The system described by Kaseje does not work perfectly. It should not be overly idealized. Kaseje himself expressed considerable concern that it will be possible to break the professional and bureaucratic inertia needed to extend and sustain the program he has described.

It is clear, however, that the resources needed to enable the family to provide effective health services to its members are very similar to those identified three decades ago by Schultz to enable peasant producers to become effective suppliers of agricultural commodities. The "high pay-off" health inputs include:

(a) The capacity of the health research community to produce the new knowledge and the materials that are appropriate to the resource and cultural endowments of rural communities.

(b) The capacity of national, regional, and local institutions to make the knowledge and the materials available to families; and

(c) The formal schooling and informal education of families, particularly mothers, to make effective use of the knowledge available to them.

The second aspect of the vision of an effective global health system is the nature of the international apparatus needed to sustain a global health research system.

In the field of agriculture, the CGIAR sponsored set of international research centers serve as leaders of applied science for the Third World and accelerators of linkages between frontier science and Third World problems. There is no comparable set of internationally supported health research centers in poor countries of the tropics. Adetokunbo Lucas (1984) notes that there are only two international centers of significant size in the field of health - the International Centre for Diarrheal Disease Research in Bangladesh, and the International Centre for Insect Physiology and Ecology in Kenya (which is concerned with entomology that is relevant to both health and agriculture).

There are strong differences of opinion within the international health community as to whether a system of international health research centers, analogous to the CGIAR centers, would be appropriate or effective.

On the one hand internationally organized efforts have the advantage of achieving a critical mass of scientists concentrating on and physically located close to high-priority problems...Internationally organized research efforts can focus on specific problems in a multidisciplinary way and demonstrate economies of scale in their operations, making them attractive to external funding. On the other hand, international center salaries are high and their activities, if not carefully targeted, can supersede rather than complement national efforts (Commission on Health Research for Development, 1950:58).

Present constraints on foreign assistance funds suggest that it would be unrealistic to expect that resources could be mobilized in the mid-1990s to support an expanded system of international health research centers in the tropics. It seems more likely that the predominant model of international collaboration in the health field will be international networks linking scientists in national institutions (both in industrialized and developing countries) in goal-oriented research programs aimed at specific health problems. A successful example of such collaboration is the Special Programme for Research and Training in Tropical Diseases (TDR), co-sponsored by UNDP, the World Bank, and WHO. Started in 1976, TDR focusses on six specific diseases, (including malaria, schistosomiasis, and leprosy). In addition to supporting research it invests approximately 25 per cent of its annual budget of $30-35 million in strengthening research capacity in developing countries.

While international networks of national centers evidently can work effectively in supporting research on particular diseases, there is one extremely important function they cannot perform. The field of health research conspicuously lacks an overview mechanism. In agriculture, the CGIAR (as distinct from the set of centers it sponsors) has built highly valuable methods for surveying the world-wide agricultural research scene in relation to the needs for research results, reviewing on-going research activities (both those of the international centers and of other institutions), and proposing changes in current research priorities and institutional arrangements including where necessary the development of new research facilities.

There is no analogous, effective, independent organization in the health field for assessing progress in research, especially on developing-country health problems, identifying neglected areas, and promoting necessary action. The result is clear. At present, of the three leading infectious disease causes of death in the world (acute respiratory infections, diarrheal diseases, and tuberculosis), only diarrhea is addressed by a major, sustained research effort. That is why the Commission on Health Research for Development came to the conclusion that "a health analogue of the CGIAR assessment and promotion structure could be of great value and should be established" (Commission on Health Research for Development, 1990:59).

In response to this challenge, the World Bank has proposed the establishment of a body to coordinate research on health problems of developing countries. Meanwhile a Geneva based Council on Health Research, funded by several donors, including Canada and Sweden, has been established to provide support for essential national health research in developing countries. Discussion has been initiated to explore the complementarity between the two initiatives (Aldhous, 1993; World Bank, 1993).

Bridging the "Island Empires"

We have argued that the "island empires" of the health and agricultural sciences can learn from one another as they strive to build global research systems that can support sustainable development. Whether they can, or even should, move beyond passive learning to active cooperation remains to be seen.

There seems little merit in any grand organizational scheme that would attempt to pull the already diverse networks of research in the respective empires under a single roof. What does seem both feasible and desirable, however, is to begin some modest effort at active bridge building.At a minimum, the principals of the three empires might agree to meet regularly - at the international, the national and the local level - in order that they and their senior staff members could get to know one another and exchange information on current activities.

At a deeper level, it is essential to realize that the global health and agricultural research systems outlined in this chapter have important common elements. The global systems outlined in this chapter can be effective only as the underlying sciences - particularly the biological and the social sciences - advance. Advances in the biological sciences and the social sciences are necessary to enlarge the world's understanding of the natural and social phenomena involved in assuring food security and health maintenance. They are also needed in order to expand the capacity to apply advances in knowledge to the national and human dimensions of development in the poor countries where most of the world's people live.

The need to enlarge scientific capacity in the poorer countries of the world should not be viewed as a burden on either the developed or developing countries. Rather it is an opportunity to multiply the intellectual talent necessary to advance knowledge relevant to the achievement of sustainable development. Completion of the development of global research systems in health and agriculture is a necessary component of a global effort to establish and mobilize the intellectual capacity and energy that will be needed to sustain development.

References

Aldhous, Peter. 1993. "World Bank Report calls for Network to Bolster Research." *Science* 261 (July), p. 135.

Allen, L., A. Chavez, and G. Pelto. The collaborative research and support program on food intake and human function: Mexico Project Final Report. (Storrs, CT: University of Connecticut, 1987.)

Asian Vegetable Research and Development Center. 1986 Progress Report. AVRDC, Shanhua, Tainan, 1988. Pp. 310-313.

Balderston, J., G. Beaton, D. Calloway, H. Horan, S. Murphy, C. Rosberg, and S. Selvin. The collaborative research and support program on food intake and human function: The Management Entity, *The Management Entity Final Report*. (Berkeley, CA: University of California, 1988.)

Baum, Warren. 1986. *Partners Against Hunger: The Consultative Group on International Agricultural Research.* Washington, DC: The World Bank.

Benor, Daniel and James Q. Harrison. 1977. *Agricultural Extension: The Training and Visit System.* Washington, DC: World Bank.

Borah, W., and S.F. Cook. The Population of Central Mexico in 1548, *Ibero-America* 43 (1960). (Berkeley: University of California Press).

Boyd Orr, John. 1966. *As I Recall.* London: MacGib and Kee.

Bradley, David. 1994. Institutional Capacity to Monitor the Interactions of Agriculture and Health Change. Agriculture, Environment and Health: Sustainable Development in the 21st Century. Vernon W. Ruttan. Ed., Minneapolis: University of Minnesota Press, pp. 308-338.

Bwibo, N. and C. Neumann. The collaborative research and support program on food intake and human function: Kenya Project Final Report. (los Angeles, CA: University of California, 1987.)

Caldwell, John C. and P. Caldwell. Is the Asian Family Planning Program Model Suited to Africa? *Studies in Family Planning*, Vol. 19, No. 1 (1988).

Calloway, D.H., S.P. Murphy, and G.H. Beaton. Food intake and human function: Across-project perspective of the collaborative research program in Egypt, Kenya and Mexico. (Berkeley, CA: University of California, 1988.)

Cleland, John and Christopher Wilson. Demand Theories of the Fertility Transition: An Iconoclastic View, *Population Studies* 41 (March 1987), pp. 5-30.

Commission on Health Research for Development. 1990. *Health Research; Essential Link to Equity in Development.* Oxford: Oxford University Press.

Conway, R. Gordon and Jules N. Petty. *Fertilizer Risks in the Developing Countries: A Review* (London: International Institute for Environment and Development, 1988).

Cook, S.F., and W. Borah. The Indian Population of Central Mexico, 1531 - 1610, *Ibero-Americana* 44. (Berkeley: University of California Press, 1960).

Cook, S.F., and L.B. Simpson. The Population of Central Mexico in the Sixteenth Century, *Ibero-Americana* 31. (Berkeley: University of California Press, 1948).

Crosby, A.W. Conquistador y Pestiliencia: The First New World Pandemic and the Fall of the Great Indian Empires, *Hispanic American Historical Review* (1967), pp. 47:321-337.

_____ Virgin Soil Epidemics as a Factor in the Aboriginal Depopulation in America, *William and Mary Quarterly* (1976), pp. 33:289-299.

Denevan, W.M. (Ed.) *The Native Population of the Americas in 1492* (Madison, WI: University of Wisconsin Press, 1976).

Eicher, Carl. 1994. Building Productive National and International Agricultural Research Systems. In Agriculture, Environment and Health: Sustainable Development in the 21st Century. Vernon W. Ruttan, Minneapolis, University of Minnesota Press, pp. 77-103.

Evenson, Robert E., Donald D. Evenson and Jonathan D. Putnam. 1987. Private sector agricultural inventions in developing countries, pp. 469-511 in Vernon W. Ruttan and Carl E. Pray (Eds.). *Policy for Agricultural Research*. Boulder: Westview Press.

Galal, O., G. Harrison, N. Jerome, A. Kirksey, The collaborative research and support program of food intake and human function: Egypt Project Final Report. (West Lafayette, IN: Purdue University, 1987.)

Gibson, C. *The Aztecs Under Spanish Rule*. (Stanford, California: Stanford University Press, 1964).

Gray, W.M., Strong Association Between West African Rainfall and U.S. Landfall of Intense Hurricanes, *Science* 249 (Sept. 14, 1990): pp. 1251-1256.

Gunatelleke, Godfrey. 1994. Health Policy for Rural Areas: Sri Lanka. In Agriculture, Environment and Health: Sustainable Development in the 21st Century. Vernon W. Ruttan, Ed., Minneapolis: University of Minnesota Press: pp. 209-235.

Hassig, R. *Trade, Tribute, and Transportation: The Sixteenth-Century Political Economy of the Valley of Mexico*. (Norman: The University of Oklahoma Press, 1985).

Hatcher, J., *Plague, Population and the English Economy* (London and Basingstoke: The Macmillian Press Ltd., 1977), pp. 1348-1530.

Johnson, D.L. and T.M. Whitmore. Population Reconstruction of the Egyptian Nile Valley: 4000 B.C. to Present. Technical Paper # 3 in 1986, *Millennial Long Waves of Human Occupance Project*, NSF, Clark University.

Judd, M. Ann, James K. Boyce and Robert E. Evenson. 1987. Investment in agricultural research and extension. *In Policy for Agricultural Research*. Vernon W. Ruttan and Carl E. Pray

(Eds.). Boulder: Westview Press: 2-38.

Kaseje, Dan C. O. 1994. Health Systems for Rural Areas: Kenya. In Agriculture, Environment and Health: Sustainable Development in the 21st Century. Vernon W. Ruttan, Ed., Minneapolis, University of Minnesota Press, pp. 236-256.

Lucas, Adetokunbo O. 1994. The Institutional Infrastructure for Health Research in Developing Countries. In Agriculture Environment and Health: Sustainable Development in the 21st Century, (Ed.) Vernon W. Ruttan. (Minneapolis: University of Minnesota Press), pp. 187-208.

Mata, Leonardo. *Children of Santa Maria Cauque*, (Cambridge: MIT Press, 1978).

Mata, Leonardo. 1988. A public health approach to the "food-malnutrition-economic recession" complex. In *Health, Nutrition, and Economic Crises: Approaches to Policy in the Third World.* David E. Bell and Michael R. Reich (Eds.), pp. 265-275. Dover: Auburn House.

McEvedy, C., and R. Jones. *Atlas of World Population History,* (New York: Penguin Books, 1978).

Meadows, D.H., D.L. Meadows, J. Randers, and W.W.I. Behrens. *The Limits to Growth*, (New York: The New American Library, Inc., 1972).

National Research Council and Institute of Medicine, *U.S. Capacity to Address Tropical Infection Disease Problems*, (Washington, DC: National Academy Press, 1987).

Parik, Kirit. 1993. Agriculture and Food System Scenarios for the 21st Century. In Agriculture, Environment and Health: Sustainable Development in the 21st Century. Ed., Minneapolis: University of Minnesota Press, pp. 26-47.

Pray, Carl E. 1987. Private Sector Agricultural Research in Asia. *In Policy for Agricultural Research.* Vernon W. Ruttan and Carl E. Pray (Eds.), pp. 411-431. Boulder: Westview Press.

Randers, J. (Ed.), *Elements of the System Dynamics Method* (Cambridge: M.I.T. Press, 1980).

Rosenzweig, Mark R., Human Capital, Population Growth and Economic Development: Beyond Correlations, *Journal of Political Economy* 10 (1988), pp. 83-111.

Russell, J.C., *British Medieval Population* (Albuquerque, N.M., University of New Mexico Press 1948).

Ruttan, Vernon W. 1982. *Agricultural Research Policy.* (Minneapolis: University of Minnesota Press 1992.)

Ruttan, Vernon W. 1986. Toward a Global Agricultural Research System: A Personal View. *Research Policy*, 15:307-327.

Ruttan, Vernon W. (Ed.) 1989. *Biological and Technical Constraints on Crop and Animal Productivity: Report on a Dialogue.* St. Paul: University of Minnesota Department of Agricultural and Applied Economics. November).

Ruttan, Vernon W. (Ed.) 1990a. *Resource and Environmental Constraints on Sustainable Growth on Agricultural Production: Report on a Dialogue.* St. Paul: University of Minnesota Department of Agricultural and Applied Economics.

Ruttan, Vernon W. (Ed.) 1990b. *Health Constraints in Agricultural Development.* St. Paul: University of Minnesota Department of Agricultural and Applied Economics.

Ruttan, Vernon W. (Ed.) 1992. *Sustainable Agriculture and the Environment: Perspectives on Growth and Constraints.* Boulder: Westview Press.

Ruttan, Vernon W. (Ed.) 1994. Agriculture, Environment and Health: Toward Sustainable Development in the 21st Century. Minneapolis: University of Minnesota Press.

Sanders, J.H., J.G. Nagy and S. Ramaswamy, Developing New Agricultural Technologies for the Sahelian Countries: The Burkina Faso Case, *Economic Development and Cultural Change* 39(1) (October 1990), pp. 1-22.

Savadogo, Kimseyinga. *Production Systems in the Southwestern Region of Burkina Faso*, mimeo prepared for INTSORMIL,

Schultz, Theodore W. 1964. *Transforming Traditional Agriculture.* New Haven: Yale University Press.

Slack, P. *The Impact of Plague in Tudor and Stuart England* (London: Routledge & Kegan Paul, 1985).

Tendler, Judith. 1994. Tales of Dissemination in Agriculture. In Agriculture, Environment and Health: Toward Sustainable Development in the 21st Century. Vernon W. Ruttan, Ed., Minneapolis: University of Minnesota Press, pp. 146-180.

Turner, B.L. II. Population Reconstruction of the Central Maya Lowlands: 1000 B.C. to Present, Technical paper #2, in *Millennial Longwaves of Human Occupance Project*, NSF, Clark University, 1986).

Turner, B.L. II, and Patricia A. Benjamin. 1993. Fragile Lands: Identification and Use for Agriculture. In Agriculture, Environment and Health: Toward Sustainable Development into the 21st Century. Ed. Vernon W. Ruttan. Minneapolis: University of Minnesota Press, pp. 104-145.

Turner, B.L. II, and S.B. Brush. *Comparative Farming Systems* (New York: The Guilford Press, 1987).

Weinberg-Andersson, S., E. Carter and C. Neumann. Unpublished data from 1984-87 field study in Embu, Kenya. UCLA, Los Angeles.

Whitmore, T.M., B.L. Turner, D.L. Johnson, R.W. Kates, and T.R. Gottschang (Eds.). Population and Environmental Transformation: A Long Term View, *The Earth as Transformed by Human Action* (Cambridge: Cambridge University Press, 1990).

Whitmore, T.M., and B.L. Turner II. Population Reconstruction of the Basin of Mexico: 1150 B.C. to Present, Technical Paper #1, *Millennial Longwaves of Human Occupance*, 1986).

World Bank. 1992. *Development and the Environment: World Development Report* 1992. Oxford: Oxford University Press.

World Bank. 1993. *Investing in Health: World Development Report, 1993.* Oxford: Oxford University Press.

Printed and bound by CPI Group (UK) Ltd, Croydon, CR0 4YY

23/10/2024

01778240-0008